BASIC LUMBER
ENGINEERING
FOR BUILDERS

by Max Schwartz

Disk Included
Wood Beam Sizing™
by NorthBridge Software Inc.

D1427459

Craftsman Book Company
6058 Corte del Cedro / P.O. Box 6500 / Carlsbad, CA 92018

The author would like to acknowledge the contributions of the following companies and organizations for furnishing materials used in this book:

Air-Nail, A Subsidiary of International Staple and Machine Co.
5335 Reisner Way
South Gate, CA 90280

APA - The Engineered Wood Association
7011 South 19th Street
Tacoma, WA 98411-0700

Buildex - Division of Illinois Tool Works, Inc.
3600-T West Lake Avenue
Glenview, IL 60025

Gurley Precision Instruments
514 Fulton Street
Troy, NY 12180-3315

International Conference of Building Officials
5360 Workman Mill Road
Whittier, CA 90601-2298

Nikon, Inc.
1901 Pennsylvania Avenue N.W.
Washington DC 20006-3405

Simpson Strong-Tie Co., Inc.
4637 Chabot Drive, Suite 200
Pleasanton, CA 94588

Southern Building Code Congress, International
900 Montclair Road
Birmingham, AL 35213

Stanley-Bostitch, Inc.
10 Briggs Drive
Department T
East Greenwich, RI 02818

All portions of the Uniform Building Code™ are reproduced from the 1997 edition, copyright © 1997, with the permission of the publisher, the International Conference of Building Officials.

Library of Congress Cataloging-in-Publication Data

Schwartz, Max 1922--
    Basic lumber engineering for builders / by Max Schwartz.
      p.   cm.
    Includes index.
    ISBN 1-57218-042-0
    1. Framing (Building) 2. Lumber -- Grading. 3. Strength of materials.   I. Title.
TH2301.S38   1997
694'.2--dc21   97-19913
                        CIP

©1997 Craftsman Book Company
Fourth printing 2002

Contents

Chapter 1

Wood Basics

Wood has always been the most important construction material used in this country. Blessed with great forests, early American pioneers depended on wood for their frontier forts, cabins, and towns.

There are still more wood-frame buildings built in America than any other type. That means just about every contractor in America needs a working knowledge of practical wood engineering and construction. And you can't just rely on what you've learned through years of experience in the construction business. The country has suffered catastrophic hurricanes, tornadoes and earthquakes in the last decade. They've led to changes in the building codes and in local construction methods in areas where these disasters are likely to occur.

The introduction of composite and prefabricated structural members has led to additional revisions in the code, as well as new methods of design and construction.

When building codes were developed in the late 1920s, the writers adapted minimum standards for the "average" type of building and occupancy. They incorporated tables that listed minimum sizes of rafters, girders and joists for various spans. Most builders didn't worry about calculating stresses or deflections. They just went "by the code," or more accurately, by minimum requirements of the code. There was little need for engineering in wood construction except for beam spans greater than 25 feet.

The purpose of this book is to fill this gap and provide basic engineering rules for typical and nontypical conditions. It explains why wood is a complex structural material that reacts to the environment. It covers both the natural wood products and the newer composite and prefabricated wood structural members. Let's start with a look at how wood grows, is made into construction lumber, and how it changes with varying environmental conditions.

This revised edition of *Basic Lumber Engineering for Builders* incorporates requirements of the *2000 International Building Code* and the supplements of 2001 and 2002.

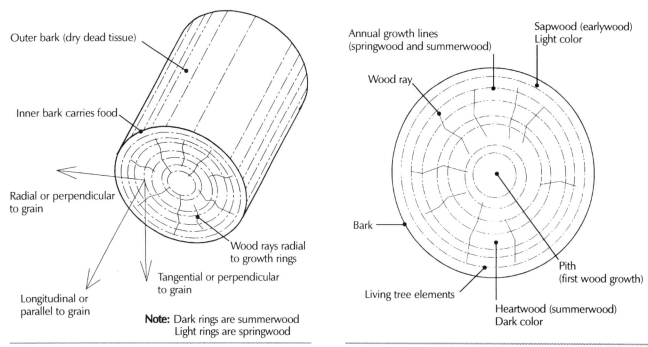

Figure 1-1 *Wood growth layers and grain direction*

Figure 1-2 *Heartwood and sapwood*

Lumber Characteristics

A tree grows by adding new layers of cells from the center (pith) outward. Every year, a new annual growth ring made up of a light layer and a dark layer is added (shown in Figures 1-1 and 1-2). The thicker light layer is called *springwood*, which is the fast growth added during spring. The thinner dark layer is called *summerwood*. It's the slow growth added during summer. Since summerwood is denser and stronger than springwood, the proportion of summerwood to springwood determines the quality and strength of the wood.

A tree trunk is made up of sapwood and heartwood. *Sapwood* is located under the bark and allows water and nourishment to pass from the root system to the leaves. It's not a good idea to use lumber cut from sapwood for construction since it contains sap. *Heartwood* is inside the sapwood and gives structural strength to the tree. Heartwood is denser than sapwood and more resistant to shrinkage, warping, and decay.

Hardwood and Softwood

Hardwood such as oak or maple comes from deciduous (broadleaf) trees while softwood such as pine or fir comes from evergreen (cone bearing) trees. Softwood is used mainly for framing lumber and plywood. Hardwood is generally used for wood furniture, decorative interior paneling, trim, and strip flooring. Most common softwoods resist splitting and have excellent nail-holding qualities.

Wood Groups

Hardwoods and softwoods are classified into four groups according to their density (specific gravity). Group I is the densest and Group IV is the least dense. Lumber groups are shown in Figure 1-3.

Figure 1-4 shows some of the physical properties of common hardwoods. The densities shown in the figure are the approximate shipping weights of air-dried lumber in pounds per cubic foot. Specific gravity is based on a 12 percent moisture content.

Redwood

Redwood is a special type of wood that's more resistant to the elements and infestation than other popular species used in construction. Although it's relatively expensive, redwood is widely used because:

▌ It shrinks very little and is resistant to warping, cupping, and nail-popping.

▌ The heartwood is naturally resistant to decay and insects.

▌ Redwood has fewer volatile resins, so it resists burning. If it ignites, it burns slowly and forms a protective char layer over the wood underneath.

▌ Redwood shrinks and swells very little during varying moisture conditions, and its unseasoned moisture content is low, ranging from 6 to 7 percent.

▌ Redwood holds paint and stains better than other softwoods.

Redwood is graded in descending order of quality and cost as:

▌ Clear all-heart vertical grain (V.G.)

▌ Clear all-heart

▌ Select heart

▌ Construction heart

▌ Utility

▌ Merchantable

Clear all-heart vertical grain is the highest redwood grade because it's all heartwood, free of knots. It's used for fine siding, paneling, cabinets, and finish carpentry. Select heart is also all heartwood, free from shakes and splits. Clear all-heart is similar to clear all-heart V.G. except that it contains some cream-colored sapwood. This grade is normally kiln-dried and used as a less expensive alternative to clear all-heart V.G. redwood.

Construction heart resists weathering and is economical. It's sold surfaced or rough, and it's normally unseasoned. It contains knots $2\frac{1}{2}$ to 4 inches in diameter and is normally used for posts, fencing, decking, and outdoor structures.

Group	Species
I	ash, beech, birch, hickory, maple, oak
II	Douglas fir-larch, southern pine, sweet gum
III	aspen, California redwood, Douglas fir - south, eastern hemlock - tamarack, eastern spruce, hem-fir, Idaho white pine, lodge pole pine, mountain hemlock, northern pine, ponderosa pine - sugar pine, red pine, Sitka spruce, southern cypress, spruce - pine - fir, western hemlock yellow poplar
IV	aspen, balsam fir, California redwood (open grain), coast Sitka spruce, and coast species, cottonwood - black and eastern, eastern white pine, Engelmann species, northern white cedar, subalpine fir, western cedars, western white pine

Figure 1-3 Wood groups

Species	Splitting resistance	Nail holding	Density (pcf)	Specific gravity (pcf)
ash	good	excellent	40.5	0.6
basswood	excellent	good	26	0.37
beech	fair	fair	46.1	0.64
birch	fair	fair	44	0.62
chestnut	fair	fair	31	0.43
cottonwood	excellent	excellent	28.5	0.4
cypress	excellent	excellent	30	0.46
elm	fair	excellent	43.2	0.5
hickory	fair	good	48.2	0.72
magnolia	good	excellent	32	0.5
maple	fair	good	44.5	0.63
oak, red	good	excellent	47.3	0.63
poplar	excellent	good	26	0.42
sycamore	excellent	excellent	35	0.49
walnut, black	fair	good	32	0.39
willow, black	fair	good	32	0.39

Figure 1-4 Some physical properties of common woods

Utility grade redwood contains sapwood and is less resistant to decay and insects. It's sold rough or surfaced and is usually unseasoned. This grade is used for subfloors, fencing, decking, and outdoor structures.

Merchantable grade is the most economical. It contains larger knots and more serious defects than are permitted in the higher grades. Both heartwood and sapwood pieces are included in this grade, as well as shakes, stains, splits, and knotholes. It's used for decking, subfloors, temporary low-cost construction, and interior work where a knot-free surface isn't required. It's sold surfaced or rough, and is usually unseasoned.

Lumber Grading

Lumber is graded according to the number, size and location of grains, knots, and checks in it. Higher grade lumber has the least amount of these defects. Lumber graders classify softwood lumber for construction purposes into three basic groups:

- Lumber for primary construction purposes
- Lumber for secondary construction, such as wales and studs used in building concrete forms
- Lumber for architectural purposes

Hardwood mills grade their lumber by the percentage of board that's free of defects. They also grade lumber by the amount of usable lumber in each piece, or the number of clear face cuttings that can be made between knots and cracks. They classify hardwood as firsts, seconds, select, No. 1 common, and No. 2 common.

Firsts is the highest grade of hardwood. Each piece must be at least 4 inches wide and 5 feet long or 3 inches wide and 7 feet long. Six-inch widths are allowed if $92^2/3$ percent of each piece is clear of knots and defects.

Seconds are usually mixed with Firsts and called *FAS* (Firsts and Seconds). At least 83 percent of each board must be clear. Each piece of Seconds must also be at least 4 inches wide and 5 feet long or 3 inches wide and 7 feet long.

Select is another rating of hardwood. Each piece of Select must also be at least 4 inches wide and 5 feet long or 3 inches wide and 7 feet long. Each piece must be $91^2/3$ percent clear.

No. 1 common is the fourth class of hardwood. Each piece of No. 1 common must be at least 4 inches wide and 2 feet long or 3 inches wide and 3 feet long.

The lowest grades include No. 2 common, sound wormy, No. 3A common, and No. 3B common. Each piece must be at least 3 inches wide and 2 feet long.

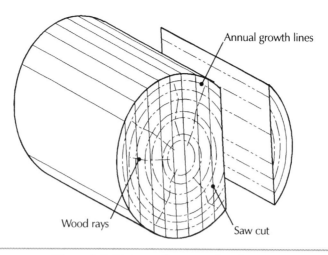

Figure 1-5 Plain- or flat-sawn lumber

Lumber Shapes

The way timber is cut into lumber affects its strength, its degree of shrinkage, and its appearance. Plane- or flat-sawn lumber is cut in slices parallel to one side of the log, as shown in Figure 1-5. Quarter-sawn lumber is made by sawing lumber perpendicular to the exterior of the log, as shown in Figure 1-6. Quarter-sawn lumber is more expensive than flat-sawn lumber because there's more waste and labor in cutting it. But quarter-sawn wood twists and cups less, holds paint better, wears more evenly, and swells and shrinks less than flat-sawn lumber. The amount that a piece of lumber will shrink and warp depends on where the piece was cut from the tree. Wood will shrink and warp perpendicular to the grain as it loses or gains moisture, but it won't shrink or swell much parallel to the grain. See Figure 1-7.

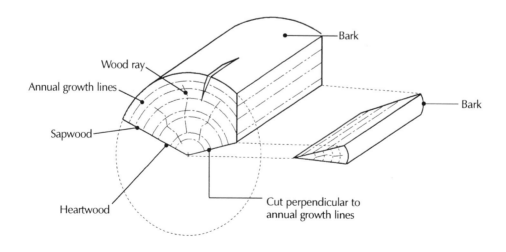

Figure 1-6 Quarter-sawn lumber

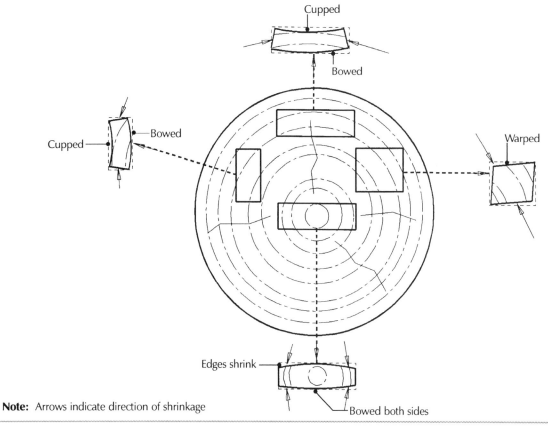

Note: Arrows indicate direction of shrinkage

Figure 1-7 *Effect of sawing on shrinkage*

Mills saw, edge, trim, and plane trees into lumber. The ultimate shape of lumber depends on its size and how it's used in construction. Some of these shapes are:

1) Boards: Lumber less than 2 inches thick and 2 or more inches wide. Boards less than 6 inches wide are called *strips*.

2) Dimension lumber: The National Grading Rule for dimension lumber classifies dimensions into width and use categories. These are:

   ■ Dimensions up to 4 inches wide are classified as structural light framing, light framing, and studs

   ■ Dimensions 5 inches and wider are classified as structural joists and planks

   ■ Appearance framing grade is 2 inches and wider and used for high strength and appearance

3) Beams and stringers: These carry loads on the narrow face of a member. Members are 5 inches or more thick, and at least 2 inches wider than they are thick.

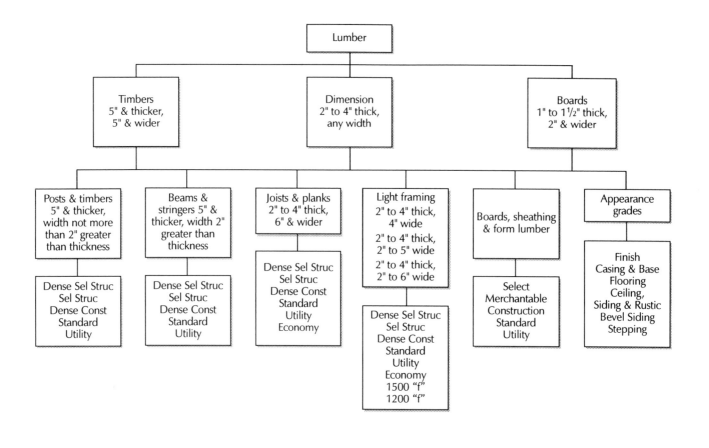

Figure 1-8 *Lumber grade chart*

4) Joists and planks: These members are similar to beams and stringers except they're 2 to 4 inches thick and at least 6 inches wide. Joists are installed to carry loads on the narrow face, while planks used for decking or flooring carry loads on the wide face.

5) Posts and timbers: This shape includes lumber with a square or almost square cross section used for posts or columns. These members are 5 inches or more thick and less than 2 inches wider than they are thick.

Figure 1-8 shows more categories of various types of lumber.

Round Sections

Lumber mills also manufacture timber with round cross sections. Round shapes are used for pilings, trestles, wharves, falsework, utility poles, and gin poles. Douglas fir and southern yellow pine are the most common species used for pilings in North America. Telephone and power companies normally use Douglas fir, southern pine, and lodgepole pine for their poles.

Lumber Dimensions and Surfaces

When the mills saw lumber into rectangular shapes, they call the narrow dimension the thickness and the wide dimension, the width. When a rectangular member is installed as a joist or rafter, with its wide dimension vertical, the width of the board is called the depth. In other words, the term *depth* is the dimension of the cross section measured parallel to the direction of the main load on the member. The term *width* is the dimension of the cross section measured perpendicular to the direction of the main load.

Lumber is usually sawn in standard sizes, lengths, widths, and thicknesses. You can identify lumber by its nominal (named) size or by its net (dressed) size. Carpenters normally refer to the nominal size. Quoted prices usually refer to rough or nominal dimensions prior to surfacing. You have to take these dimensional lumber differences into consideration when you make a design. A typical example of dimensional difference is the common 2 × 4. Previously, the actual dressed size of a rough or nominal 2 × 4 was $1^5/_8$ by $3^5/_8$ inches. The present dressed size is $1^1/_2$ by $3^1/_2$ inches.

Some mills describe dressed lumber by the number of sides and edges that are surfaced. Here's some examples:

Symbol	Meaning
S1S	Surfaced one side
S2S	Surfaced two sides
S4S	Surfaced four sides
S1S1E	Surfaced one side, one edge
S1S2E	Surfaced one side, two edges

In most cases, lumber suppliers deliver lumber that's surfaced on four sides, or S4S. Figure 1-9 lists nominal size, dressed size, and end area of wood members ranging from 2 × 4s to 6 × 14s.

Lumber Seasoning

The amount of moisture retained in lumber is important in construction because too much moisture can cause problems, as we will see later in this chapter. Moisture content is expressed as a percentage of the weight of water in the wood compared with the oven-dry weight of the wood. When lumber is first cut, its moisture content may be as high as 50 percent. This unseasoned lumber contains too much moisture and should be seasoned by drying before you use it.

Nominal size (in)	Dressed size (in)	Area of section (in²)	Moment of inertia (in⁴)	Section modulus (in³)
2 × 4	1½ × 3½	5.25	5.359	3.063
2 × 6	1½ × 5½	8.25	20.797	7.563
2 × 8	1½ × 7¼	10.875	47.635	13.141
2 × 10	1½ × 9¼	13.875	98.932	21.391
2 × 12	1½ × 11¼	16.875	177.979	31.641
2 × 14	1½ × 13¼	19.875	290.775	43.891
3 × 4	2½ × 3½	8.75	8.932	5.104
3 × 6	2½ × 5½	13.75	34.661	12.604
3 × 8	2½ × 7½	18.125	79.391	21.901
3 × 10	2½ × 9½	23.125	164.886	35.651
3 × 12	2½ × 11½	28.125	296.631	52.734
3 × 14	2½ × 13½	33.125	484.625	75.151
3 × 16	2½ × 15½	38.125	738.87	96.901
4 × 4	3½ × 3½	12.25	12.505	7.146
4 × 6	3½ × 5½	19.25	48.526	17.646
4 × 8	3½ × 7½	25.375	111.146	30.661
4 × 10	3½ × 9½	32.375	230.84	49.911
4 × 12	3½ × 11½	39.375	415.283	73.828
4 × 14	3½ × 13½	46.375	678.475	102.411
4 × 16	3½ × 15½	53.375	1034.418	135.661
6 × 6	5½ × 5½	30.25	76.255	27.729
6 × 8	5½ × 7½	41.25	193.359	51.563
6 × 10	5½ × 9½	52.25	392.963	82.729
6 × 12	5½ × 11½	63.25	697.065	121.229
6 × 14	5½ × 13½	74.25	1127.672	167.063
6 × 16	5½ × 15½	85.25	1706.726	220.229
6 × 18	5½ × 17½	96.25	2456.38	280.729
6 × 20	5½ × 19½	107.25	3398.484	348.563

Note: Properties are based on minimum dressed green size which is ½ inch off nominal in both b and d dimensions. Moment of Inertia and Section Modulus taken about the x-axis. For lumber surfaced 1⅝ inches thick, instead of 1½ inches, the area, moment of inertia and section modulus about the x-axis may be increased by 8.33%.

Figure 1-9 Properties of sawn lumber

Lumber is dried using air or a kiln. To air dry lumber, stack it in covered piles, separating each layer with 1-inch-thick wood strips to allow air to circulate between the layers. Let the lumber dry for three or four months, or until its moisture content isn't more than 19 percent. The moisture content of lumber can't go lower than the humidity of the surrounding air. This condition is called *equilibrium*, and it's the point where swelling or shrinking is at a minimum.

To kiln dry lumber, logs or lumber are placed into kilns (ovens) at temperatures usually between 110 and 180 degrees F, although some kilns operate at over 212 degrees F. Lumber is usually dried in a kiln for four to ten days.

Here are some advantages of using seasoned lumber:

▮ Increased strength and stiffness

▮ Reduced shrinkage, checking, and cracking

▮ Reduced weight

▮ Less susceptibility to staining, insects, and decay

▮ Reduced nail popping

▮ Better glue bonding, painting, and staining properties

▮ Better appearance

Preserving and Storing Lumber

Wood Preservatives

Two types of wood preservatives are water-borne and oil-borne preservatives. Water-borne preservatives include acid copper chromate (ACC), ammoniacal copper arsenate (ACA), ammoniacal copper zinc arsenate (ACZA), chromated zinc chloride (CZC) and chromated copper arsenate (CCA). Oil-borne preservatives include creosote, creosote-petroleum, creosote solutions and pentachlorophenol.

Lumber treated with a preservative is usually stamped to identify the type of preservative used. The stamp contains the following symbols or information:

▮ Identifying symbol, logo, or name of accredited agency

▮ Year of treatment

▮ Type of preservative used

▮ Preservative retention (amount of preservative that remains in the cell structure after the pressure process is completed)

■ Exposure category, including above ground, ground contact with fresh water, wood foundation, and salt water

■ Plant name and location

■ Moisture content after treatment

Use pressure-treated or naturally durable wood in the following locations:

■ Floor joists installed within 18 inches of exposed soil

■ Floor girders installed within 12 inches of exposed soil

■ Plates, sills, and sleepers installed on concrete or masonry in direct contact with the soil

■ Any member installed within 6 inches of exposed soil

■ Members that support moisture-permeable roofs or floors exposed to weather, unless the members are separated from the soil by sheet metal or an impervious moisture barrier

■ Framing members or sheathing in foundation walls when the wood is within 8 inches of the soil

Use aluminum, galvanized or stainless steel fasteners and connections with treated wood. Unprotected steel nails and fasteners will rust, causing unsightly stains, and they'll eventually fail. Unprotected steel anchor bolts holding mud sills to foundation walls can rust and expand, causing the concrete wall to crack and spall. Use hot-dip galvanized anchor bolts to prevent corrosion. Electroplated galvanizing is not recommended.

Storing Lumber

The framing contractor is responsible for properly storing and protecting lumber at the job site. Use the following rules when you store lumber:

■ Unload lumber in a dry place.

■ Don't store lumber in direct contact with the ground. Instead, stack the lumber on stringers to allow air circulation.

■ Store lumber under a roof if possible. Cover lumber stored in open areas with a tarp or other material that protects it from the elements, but lets moisture escape. Don't use polyethylene sheets.

■ Order lumber in the sequence you're going to use it. For example, don't order rafter material before floor joists.

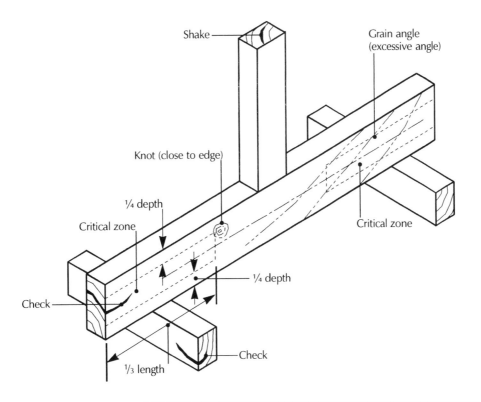

Figure 1-10 *Wood defects*

Defective Materials

A *shake* is a lengthwise separation of the wood which usually occurs between or through the rings of annual growth. A $^1/_{32}$-inch separation is called a *light shake*. A *medium shake* is $^1/_{32}$ to $^1/_8$ inch wide. A separation on only one face is called a *surface shake*. A *through shake* goes from one surface to the opposite or adjacent surface. Shakes are usually formed in a standing tree or when the tree is felled.

A *check* is a separation of wood normally occurring across or through the rings of annual growth. It's usually a result of seasoning. A *surface check* occurs on one side of a piece, and a *through check* goes from one surface to the opposite or adjacent surface.

A *split* is a separation of wood due to the tearing of wood cells. The length of a *very short split* is less than one-half the width of the piece. The length of a *medium split* is equal to twice the width of the piece. A *long split* is longer than a medium split. See Figure 1-10.

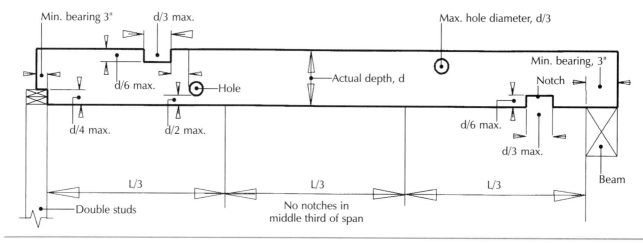

Figure 1-11 Holes and notches

Don't use defective lumber. Here's a checklist of items that will help you avoid problems when you select lumber:

1) Don't install a lower grade of lumber than was specified.

2) Be selective. Good quality lumber has small, tight knots confined to the center half of a member's cross section and within the center third of the length of the piece. See Figures 1-11 and 1-12. Loose knots at the bottom of a beam, joist, or rafter weaken the member. This is because maximum tension stresses occur at the bottom of a simple beam near the center of the span. Splits, knot holes, and other defects in this area weaken the beam.

3) Shakes, checks, and splits should be confined to the middle half of the height of a piece. This is because the maximum shear stresses occur at the beam ends adjacent to the supports. The critical areas for shear stresses are the outer thirds of the beam span and the center half of the beam depth. Defects such as checks, splits, and shakes in these areas tend to weaken a beam.

4) Checks near the ends of a member tend to weaken the member.

5) The critical end distance should be no less than three times the depth of the member. See Figure 1-12.

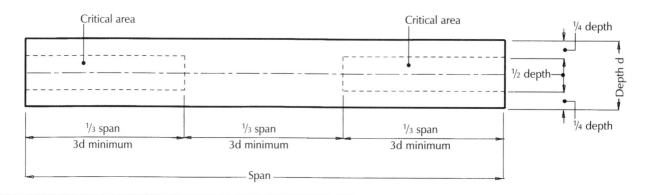

Figure 1-12 Critical areas of a wood beam

6) The slope of the grain is very important on horizontal members because the greater the slope, the weaker the member. Slope is the deviation of the wood fibers from a line parallel to the edges of a piece. It's given as a ratio of 1 in 8, 1 in 10, 1 in 12, or 1 in 15. For beams and stringers, the slope of the grain in the middle third of the board shouldn't exceed 1 in 15, with the balance of the piece 1 in 12. See Figure 1-10.

Watch for defective lumber due to fungi, rot, decay, or weathering. Fungi grow from spores which break down wood. Dry rot is a dry, crumbly, easy-to-crush condition. The term dry rot is really a misnomer, as it requires moisture to develop and should actually be called decay. Decay is the disintegration of wood due to the action of wood-destroying fungi. Weathering is the mechanical or chemical disintegration and discoloration of a wood surface caused by light, dust, sand, and a variation in moisture content.

Deterioration of wood by mold

Wood-frame buildings can be in danger of collapse from the action of fungus and mold. In such cases, the builder needs to have a basic understanding of fungus and molds:

Principal fungus stains are called "sap stains" or "blue stain."

Decay-producing fungus may, under certain conditions, attack either the sapwood or heartwood of lumber. This condition is called "decay," "rot," "dote," or "doze."

Fresh surface growth of decay fungus may appear as fan-shaped patches, strands, or root-like structures, usually white or brown. Some fruiting types of fungus take the form of toadstools, brackets, or crusts.

Fungus, in the form of microscopic, threadlike strands, permeates the wood and uses part of it as food. Some fungi live largely in the cellulose, others on the lignin, as well as the cellulose.

Other decay fungus attack the heartwood, which causes "heartrot," and rarely sapwood of a living tree, whereas some confine their activities to logs or manufactured lumber. Most of the tree-attacking fungi cease their activity after the trees have been cut down.

One type of fungus is called "Dry Rot" fungi, or "Water-conducting fungi." This type has water-conducting strands that are capable of carrying water, usually from the soil, into the building. This moistens and rots wood that would otherwise be dry.

Soft rot fungus is a type of decay caused by a fungi related to molds rather than to those responsible for brown and white rot. Soft rot is usually shallow. The affected wood is greatly degraded and often soft when wet, but immediately below that zone of rot the wood may be firm. Because soft rot is shallow, it's most likely to affect thin pieces of wood like slats or wood lath.

How to Reduce Damage by Mold and Fungi

Build with dry lumber only, free of signs of decay and not exceeding the amounts of mold and blue stain permitted by standard lumber grading rules.

Use a building design that will keep the wood dry and accelerate rain runoff.

For parts exposed to aboveground decay-producing hazards, use heartwood of decay-resistant species, or wood treated with preservatives.

For high hazard situations associated with ground contact, use pressure-treated wood.

Manufactured Wood Products

Engineered Wood Products and Structural Members

Chips and shavings from softwood cutting are used to manufacture engineered products such as pressed board, fiberboard, mineral fiberboard, particleboard, hardboard, Sturd-I-Floor members, and solid core panels.

Wood structural members can be fabricated in factories or on the job site. Some of these prefabricated structural elements are:

- Gang-nail wood trusses
- Glu-lam structural beams, posts, and arches
- Solid beams laminated on edge
- Solid beams laminated on flat
- Plywood web beams

- Stressed-skin trusses
- Gussetted trusses
- Truss joists
- Plywood box beams
- Stressed skin panels
- Wood I-beams (WI)

Prefabricated wood assemblies have several advantages:

- They're capable of long, clear spans.
- They eliminate the need for interior bearing walls and posts.
- There's no waste due to field cuts.
- Camber to compensate for deflection can be built into the member.

- They're less prone to shrinkage, warping, or twisting.
- There's less engineering design required.
- They're lightweight and easy to install.

Composite Members

There's a new generation of construction material made of plastic and waste wood. This composite material is produced in rectangular and round shapes similar to natural wood. These shapes include 1 × 6 planks, 2 × 4 to 2 × 10 beams, 4 × 4 to 8 × 8 posts, and 3- to 6-inch diameter rounds. These members are usually used for decking, railings, and light ornamental structures, but you can use some of them for primary load-bearing members.

Composite construction material has several advantages over natural materials. It needs less maintenance and it doesn't splinter, crack, warp, or decay. It's resistant to ultraviolet radiation damage. And finally, termites don't like it.

But there's also a disadvantage. Composite materials can be difficult to nail because they're very dense. You can hand nail the material as long as you hold the nail until you drive it in $^1/_2$ to $^3/_4$ inch. If you use a nail gun, set the air pressure to at least 110 psi. Common nails don't work well in this material. Instead, use ring-shank or spiral-shank nails. Predrill holes for screws or lag bolts.

Plywood Products

Plywood (structural wood panels) is generally classified as exterior and interior grade. Exterior grade plywood may be classified as Sheathing, Structural I, or Structural II. Panels are also classified as Exposure 1, Exposure 2, and Interior plywood. The panel veneers may be sanded or sand-touched. Exterior panels are manufactured with a fully waterproof glue bond and are installed in locations permanently exposed to moisture or weather.

Use Exposure 1 panels when you anticipate long construction delays before the panels can be protected. These panels have a waterproof bond. Don't substitute Exposure 1 panels in high moisture conditions when exterior-grade panels are specified. Use Exposure 2 panels in protected construction where there's only moderate moisture and few construction delays. These panels are also called interior plywood with intermediate glue.

Use exterior plywood in areas continuously exposed to moisture or weather. These panels have 100 percent waterproof glue line and no veneer below grade C.

Interior plywood is classified as Exposure I and Exposure II Sheathing, Structural I and Structural II Sheathing - Exposure I. Interior plywood is also manufactured with a sanded or sand-touched finish. Use interior panels only in a protected interior environment.

Each type of plywood panel is also classified according to the grade of its face ply, back ply, and inner plies. See Figure 1-13. Ply grades are identified by the letters A, B, C, or D. Veneer ply grades are:

A   Smooth, paintable and not more than 18 neatly-made repairs per panel

B   Solid surface with shims, sled or router repair, and tight knots extending less than 1 inch across the grain

C   Tight knots under $1^1/_2$ inches. Knot holes extending less than 1 inch across the grain. Discoloration and sanding defects that don't impair strength. Limited splits.

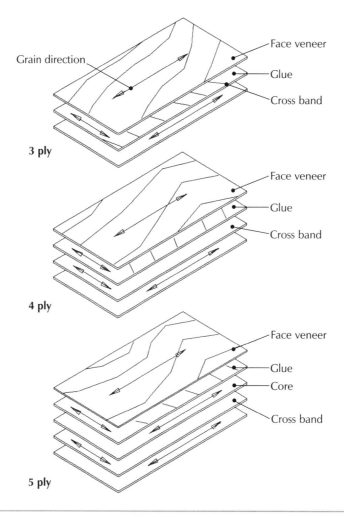

Figure 1-13 *Plywood*

C *(Improved grade)* Plugged, improved C veneer with splits limited to $^1/_8$-inch wide, and knot holes or other open defects limited to $^1/_4$ by $^1/_2$ inch. Repairs are permitted.

D Knots and knot holes extending to $2^1/_2$ inches across the grain. Limited to Exposure 1 or interior panels.

All interior grade panels have D-grade inner plies. Exterior grade panels have C-grade inner plies except for the A-C grade which has D-grade inner plies.

Plywood Span Rating

Plywood panels are stamped to show their span rating. The span rating is designated with two numbers separated by a slash (for example, 32/24). The first number is the maximum recommended spacing of supports when the panel is used for roof sheathing when the long panel dimension is installed across three or more supports. The second number is the maximum recommended spacing of supports when the panel is used for subflooring with the long panel dimension installed across three or more supports. Only one number is used to designate the span rating for wall sheathing.

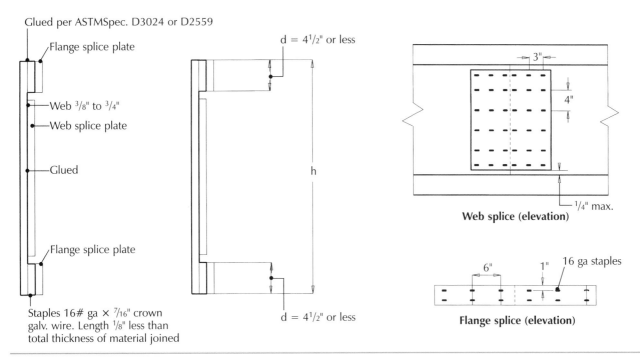

Figure 1-14 *All-plywood beam*

All-Plywood Beams

Another type of wood I-beam is called an all-plywood beam. Figure 1-14 shows the cross section and how the plywood web is spliced. All-plywood beams are fabricated with glue and staples. They can be shop- or job-fabricated. The plywood face grains for the webs, flanges, web splices, and stiffeners are oriented parallel to the span, or horizontally. The adhesive is spread uniformly over the full contact areas of the web and flanges, and the glued parts are then clamped under pressure.

Plywood Sandwich Panels

Figure 1-15 shows a cross section of a plywood sandwich panel with a skin made of plywood and a core made of urethane or similar material. Plywood sandwich panels are used for decking, rigid roofing panels, and structural insulated roof and wall panels.

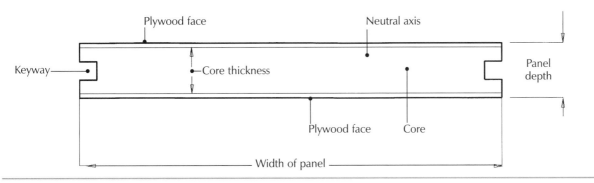

Figure 1-15 *Plywood sandwich panel*

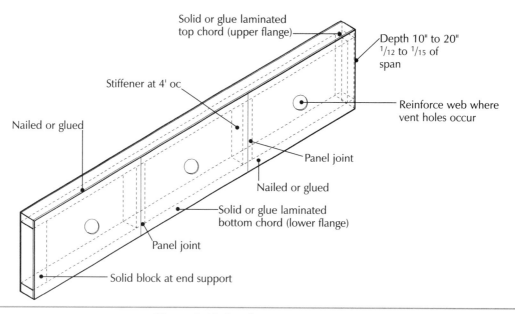

Figure 1-16 *Box beam*

Plywood Box Beams

A plywood box beam is a frame made with 2 × 4 or 2 × 6 top and bottom flanges, 2 × 4 spacers and plywood skin on each side. See Figure 1-16. Figure 1-17 shows various cross sections of plywood lumber beams. These may have one, two, or three webs. The strength of a beam depends on how well the plywood skin holds to the flanges. The plywood can be attached with nails, screws, or glue, or a combination of these. The type of glue is very important because some glues require great pressure and others require special temperature control. Figure 1-18 shows how to use a lumber beam as a garage door header.

Plywood box beams are used for girders, floor beams, or lintels over garage doors. A typical 16-foot garage door box beam lintel is 18 inches deep with single 2 × 4 top and bottom flanges covered with ³/₄-inch plywood webs. A 16-inch-deep box floor

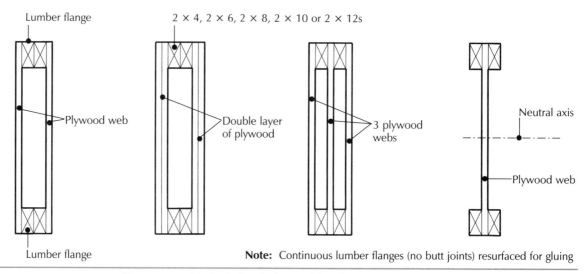

Figure 1-17 *Typical glued plywood lumber beams*

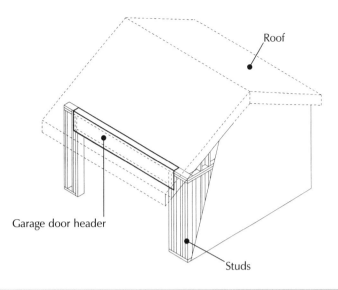

Figure 1-18 *Lumber beam garage door header*

beam with single 2 × 4 top and bottom flanges covered with $^5/_{16}$-inch plywood webs installed on 4-foot centers can span 17 feet.

The depth of a box beam is usually $^1/_{12}$ to $^1/_{15}$ of its span. Build a girder whose depth is less than $^1/_{20}$ of its span out of larger members to reduce flange stress and eliminate excessive deflection.

Box beams have both advantages and disadvantages over solid beams. The advantages are:

❙ They're much lighter than solid-sawn beams.

❙ Most of the bending stress is in the upper and lower flanges. The cross sections of these flanges can be increased to withstand increased stresses. This doesn't increase the depth of the beam.

❙ The webs resist shear stress.

❙ They're good for long spans.

The disadvantages are:

❙ The top flange must have adequate lateral support to prevent buckling.

❙ Webs may buckle if there aren't enough web stiffeners.

❙ Mildew may occur if vent holes aren't cut into the webs. (Reinforce all holes by gluing square pieces around the vent holes.)

Stressed-Skin Panels

Stressed-skin (built-up) panels are prefabricated units that can be used for roof, floor, or wall construction. Each panel is made of parallel beams separated by blocking. The frames are usually 4 feet wide and 8 to 12 feet long. The frame is covered with plywood glued to one or both sides of the framing members. The strength of the panel comes from the plywood skins which resist stress on the panel.

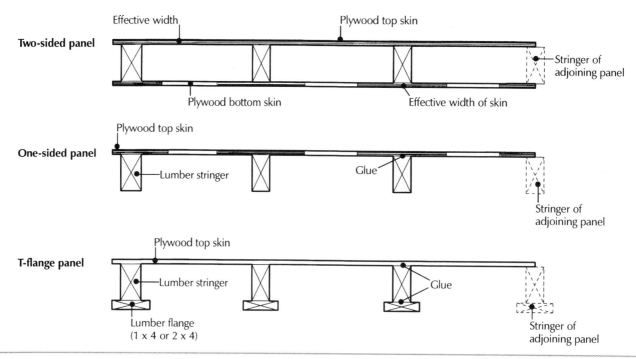

Figure 1-19 *Plywood stressed-skin panels*

Figure 1-19 shows three types of stressed-skin panels. The upper cross section shows a top and bottom skin. The center section shows a stressed-skin panel with only a plywood top skin. The lower section has a plywood top skin and a 1 × 4 or 2 × 4 lower flange glued to each beam.

Glu-lam Products

Glu-lam Structural Beams

Glu-lam beams are manufactured by specialized fabricators. These beams are made of $1^5/8$-inch thick laminations glued together and squeezed under great pressure to form a single member. See Figure 1-20. After the glue has cured, the beam is planed and sanded to its final size and finish. After curing, these beams contain only 12 percent moisture.

Most glu-lam beams are manufactured in nominal widths ranging from 3 to 16 inches and nominal depths ranging from 7 to 52 inches. The net dimensions range from $^3/4$ to $1^1/2$ inches less than nominal dimensions due to finishing and shrinkage. You can fabricate glu-lam beams on the job site, but you need strong clamps to get complete contact between the faces of the pieces. You can also use nails to help hold the pieces together, and then cover or conceal them later.

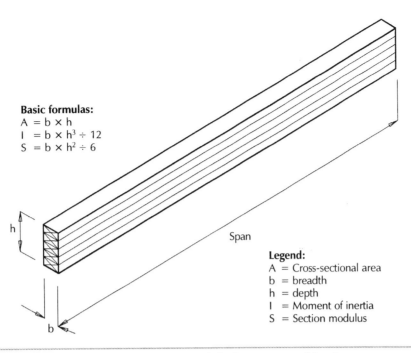

Basic formulas:
A = b × h
I = b × h³ ÷ 12
S = b × h² ÷ 6

h

Span

b

Legend:
A = Cross-sectional area
b = breadth
h = depth
I = Moment of inertia
S = Section modulus

Figure 1-20 *Properties of glu-lam structural lumber*

Glu-lam beams may be straight single-tapered for shed-type roofs, or straight double-tapered for roofs with a ridge at the center of the span. Both designs provide slope for drainage to the building edge. See Figure 1-21.

Glu-lam beams come in several grades of finish, including Premium Appearance, Architectural Appearance, and Industrial Appearance. These grades apply to the timber surfaces and include growth characteristics, inserts, wood fillers, and surfacing operation. They don't apply to the laminating procedure, design values, or grades of lumber used. Premium Appearance grade is the highest quality and the most expensive. It's used in the interiors of auditoriums, churches and other assembly buildings. Use Architectural Appearance grade for exposed installations where appearance is important. Use Industrial Appearance grade for industrial plants, warehouses, and garages where appearance isn't important.

You can order glu-lam timbers with various types of finishes and coating. Some of these are:

▌ Not sealed

▌ Sealed with factory-applied clear sealer

▌ Sealed with factory-applied white lead
    and oil primer

▌ Sealed with one coat of factory-applied TSI sealer-stain

▌ Sealed with a special type of sealer

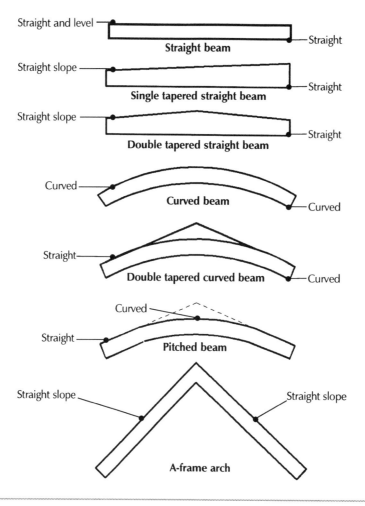

Figure 1-21 *Types of glu-lam beams and arches*

Straight glu-lam timbers may also be delivered with various types of protections:

▌ Not wrapped if shipped in closed railroad cars

▌ Load wrapped if shipped in open railroad cars or trucks

▌ Bundle wrapped

▌ Individually wrapped

You can use glu-lam beams for many residential building members such as ridge beams, vaulted ceiling beams, garage door headers, basement beams, and window headers. Since the maximum bending stresses are at the top and bottom of any beam, use a stronger grade of lumber for the top and bottom plies. You can use a lower grade of lumber for the center plies of a beam.

Figure 1-22 shows the structural properties for most sizes of glu-lam beams.

Nominal width (in)	No. of 1⁵⁄₈″ laminations	Net finish size (in)	Area (in²)	Section modulus (in³)
2	4	$2^1/_4 \times 6^1/_2$	14.6	15.8
3	4	$3^1/_4 \times 6^1/_2$	21.1	22.9
4	5	$3^1/_4 \times 8^1/_8$	18.3	24.8
5	4	$4^1/_4 \times 6^1/_2$	27.6	29.9
6	4	$5 \times 6^1/_2$	32.5	35.2
3	6	$2^1/_4 \times 9^3/_4$	21.9	35.7
4	5	$2^3/_4 \times 8^1/_8$	26.4	35.8
6	4	$5^1/_4 \times 6^1/_2$	34.1	37
5	5	$4^1/_4 \times 8^1/_8$	34.5	46.8
3	7	$2^1/_4 \times 11^3/_8$	25.6	48.5
4	6	$3^1/_4 \times 9^3/_4$	31.7	51.5
6	5	$5 \times 8^1/_8$	40.6	55
6	5	$5^1/_4 \times 8^1/_8$	42.7	57.8
5	6	$4^1/_4 \times 9^3/_4$	41.4	67.3
4	7	$3^3/_4 \times 11^3/_8$	37	70.1
8	5	$7 \times 8^1/_8$	56.9	77
6	6	$5 \times 9^3/_4$	48.8	79.2
6	6	$5^1/_4 \times 9^3/_4$	51.2	83.2
4	8	$3^1/_4 \times 13$	42.3	91.6
5	7	$4^1/_4 \times 11^3/_8$	48.3	91.7

Figure 1-22 *Properties of glu-lam lumber*

Glu-lam Posts and Arches

Glu-lam posts are popular because they're straight, stable, and can carry a heavy load. Also, they're less likely to warp, twist, or crack than solid-sawn wood posts.

Some glu-lam beams are manufactured curved throughout, or with a combination straight top and curved bottom. See Figure 1-21. The curved bottom provides additional headroom under the ceiling. The curved beam has a concentric curve top and bottom with the entire arch of equal depth. The double-tapered curved beam has a straight laminated built-up portion on the top surface which forms a peak. The pitched beam has straight tangent sections at the outer ends of the curved beam.

Curved glu-lam arches shown in Figure 1-23 are manufactured as:

▌ Radial or full arch. A full arch has no hinges and is installed so that one end is free to adjust itself to horizontal displacement.

▌ Gothic arch, with hinges at the supports and at the peak.

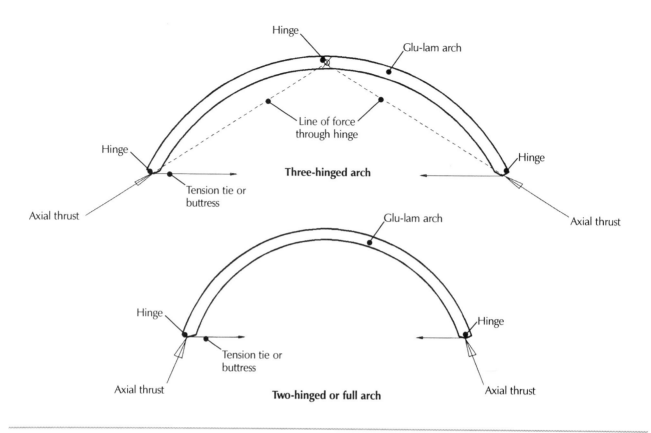

Figure 1-23 Glu-lam arches

- Two-hinged or constant radius arch. It's strengthened horizontally by the tie rod or buttress foundation.

- Three-hinged arch, which is also called a three-hinged barrel arch or Tudor arch. The three-hinged arch has a hinge at each support and a third at the crown of the arch. The arch is strengthened horizontally by a steel tie rod installed under the slab which extends across the building to the opposite support. An alternative to the tie rod is a buttress foundation which can resist the horizontal thrust of the arch. This type of arch is relatively easy to calculate since all of the reactions pass through the hinges. It's the most commonly-used arch.

- Parabolic arch. This type is not common, but it's structurally efficient.

- A-frame or teepee arch is commonly used for bulk storage buildings and recreation homes.

The arch shape provides the maximum floor space for very large buildings such as auditoriums, stadiums, and airplane hangers. You can saw the top surface of a glu-lam arch. Then you 'taper' and 'pitch' it to an unsawn sloped surface.

Glu-lam Shop Drawings

Glu-lam wood fabricators will normally furnish shop drawings to the contractor for approval before they fabricate the members. When you review the drawings, make sure the fabricator certifies that the material, manufacture, and quality control conform with ANSI/AITC A-190.1-1992 or later, and that the correct finish is specified. Verify that the design values for the top and bottom laminations are at least equal to the required design values. Also make sure the moisture content is correct and verify whether the adhesive used is for dry or wet service. Wood with less than 16 percent moisture content is considered dry and over 16 percent is considered wet.

LVL Members

Vertically laminated lumber (LVL) members are engineered wood products that are used for joists, beams, and headers. LVL joists are manufactured $7^{1}/_{4}$ to 18 inches deep and $1^{1}/_{2}$ to $3^{1}/_{2}$ inches thick. LVL beams are available in sizes from $1^{3}/_{4}$ to $5^{1}/_{4}$ inches wide and $9^{1}/_{4}$ to 18 inches deep.

LPI-Joists and LPI-Beams

LPI-joists and LPI-beams are engineered wood I-beams used as an alternative to solid-sawn beams and joists. (LPI is the brand name of engineered wood I-beams manufactured by Louisiana-Pacific. Similar beams are made by other manufacturers.) These members have a more uniform strength, stiffness, size, weight, and appearance than solid-sawn members. They also have less shrinkage, twisting, warping, and crowning. See Figure 1-24.

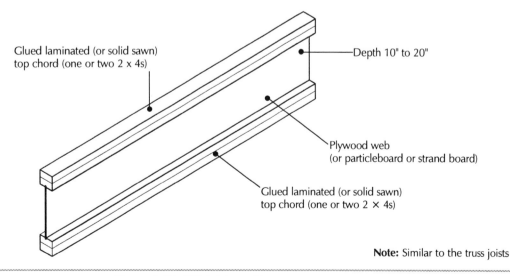

Glued laminated (or solid sawn) top chord (one or two 2 x 4s)

Depth 10" to 20"

Plywood web (or particleboard or strand board)

Glued laminated (or solid sawn) top chord (one or two 2 × 4s)

Note: Similar to the truss joists

Figure 1-24 LPI-joists and beams

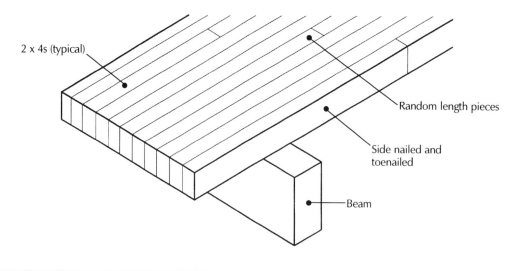

2 x 4s (typical)

Random length pieces

Side nailed and toenailed

Beam

Figure 1-25 *Wood laminated deck*

LPI-joists are made with an oriented strand board web notched into top and bottom wood flanges. The flanges are made of kiln-dried solid-sawn lumber $1^{1}/_{2}$ to $2^{1}/_{4}$ inches wide. The full depth of these LPI-joists ranges from $9^{1}/_{2}$ to 24 inches. They are manufactured up to 48 feet long.

LPI-beams are similar to the LPI-joists with depths ranging from $9^{1}/_{4}$ to 24 inches. Webs are made of plywood or OSB. Flanges are either solid-sawn wood or LVL shapes and are 4×2, 3×2, or 2×2. For the same carrying capacity, these joists use 55 percent less lumber than a solid-sawn joist.

Wood Decking Laminated on Edge

One of the best types of flooring made for heavy loading is decking laminated on edge. Use 2×4s on edge fastened with $3^{1}/_{2}$-inch coated nails on 24-inch centers. Toenail each piece through the tongue and face-nail each piece at every support. Install the boards so that the butts are staggered throughout the deck. Since the 2×4s are spliced at different locations, the decking acts as a continuous beam. This reduces deflection (or sag) and allows a heavier load. See Figure 1-25.

Job-Fabricated Beams Laminated on Edge

You can make larger beams from smaller members by laminating two or more beams on edge. You can fabricate this type of beam at the job site using nails, bolts, split rings, and glue to hold the members together. This is a cost-efficient and time-saving procedure you can use when a large solid member isn't available at the lumber yard. It's also something you can do indoors to keep crews working during bad weather.

Wood Trusses

The main advantage of trusses is that they require no intermediate support, so they provide a clear span between exterior walls. As a result, they're widely used for buildings with large open areas such as markets, factories, and assembly rooms. They also allow flexibility in interior design since there are no interior load-bearing partitions. Their open webs allow ample room for plumbing, conduits, and ductwork in the ceiling. Ceiling joists can be attached to the lower chord to provide a smooth ceiling. In buildings such as residences, where the joists are spaced no more than 24 inches on center, the bottom chords usually serve as ceiling joists.

Most wood trusses are engineered and fabricated off-site in an authorized fabricating shop and delivered to the site fully assembled and ready for installation. An approved fabricator is periodically inspected by the building department. The fabricator must submit a certificate of compliance that the work was performed according to the approved plans and specifications of the truss. The certificate is submitted to the building official and to the engineer.

Trusses consist of three major parts, including the top chord, bottom chord, and web members. The chords are usually made of 2 × 4 or 2 × 6 solid-sawn or glu-lam lumber. The top chord of a typical truss is made of select structural Douglas fir and the lower chord is made of industrial grade Douglas fir. Web members can be made from lower grade lumber such as No. 2 Douglas fir or southern pine.

The type of truss, size, and arrangement of members depends on the shape of the structure, the total roof loads, and the amount of clear floor space required. The top and bottom chords and interior web members can be solid wood sections or laminated from multiple sections. Laminated members can be glued or mechanically laminated. Steel rods or other steel shapes can be used for tension members in timber trusses. Many common types of trusses are shown in Figure 1-26.

Figure 1-27 shows six multispan buildings using trusses and tapered glu-lam girders. Multispan trusses have specific advantages:

■ They're less costly than several single-span trusses or girders.

■ They're not as deep as single-span trusses or girders.

■ The cantilever over interior supports reduces bending stresses, which reduces the required depth of the section.

Of course, they also have disadvantages:

■ They require interior columns and footings.

■ More erection labor is required for multiple spans than for a single span.

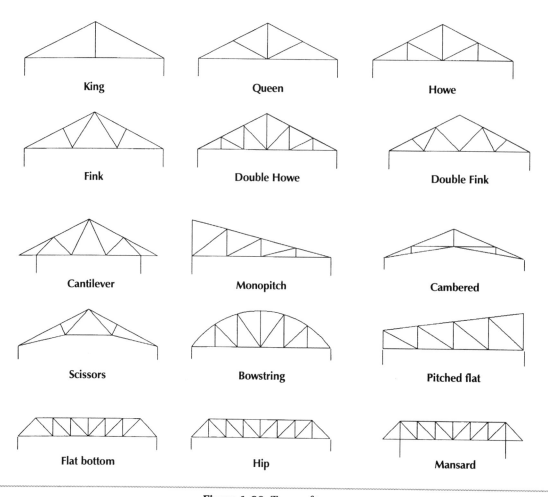

Figure 1-26 *Types of trusses*

A saw-tooth truss roof allows natural light and ventilation in a building when the vertical portions have louvered vents and glazed windows. This helps save electricity during the day. Face the glazed areas north to avoid direct sunlight into a building. A gussetted wood truss is similar to a standard wood truss except the web members are attached to the top and bottom chord by $5/16$-inch-thick plywood gussets.

Glu-lam girders cost less than trusses and reduce the volume of a building that has to be heated or cooled. Glu-lam girders may be installed as a simple span or a cantilever (Figure 1-27).

Pitched trusses have slopes normally ranging from 3 in 12 to 5 in 12, but you can use slopes up through 12 in 12. Most trusses are built with a lower chord camber of $1/360$ of the span to compensate for the deflection of the truss under load.

Connections between web members and the chords may be bolted or attached with 20-gauge galvanized steel gang-nails. These plates are placed on both faces of the truss. Each plate has as many as four teeth per square inch and provides a rigid connection between truss members. See Figure 1-28.

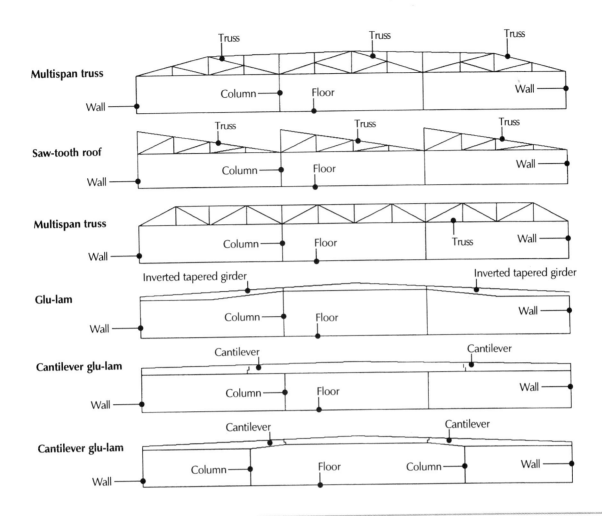

Multispan truss

Truss　　　Truss　　　Truss

Wall　Column　Floor　Wall

Saw-tooth roof

Truss　　Truss　　Truss

Wall　Column　Floor　Wall

Multispan truss

Truss

Wall　Column　Floor　Wall

Glu-lam

Inverted tapered girder　　Inverted tapered girder

Wall　Column　Floor　Wall

Cantilever glu-lam

Cantilever　　　Cantilever

Wall　Column　Floor　Wall

Cantilever glu-lam

Cantilever　　　Cantilever

Wall　Column　Floor　Column　Wall

Figure 1-27 Multiple span roofs

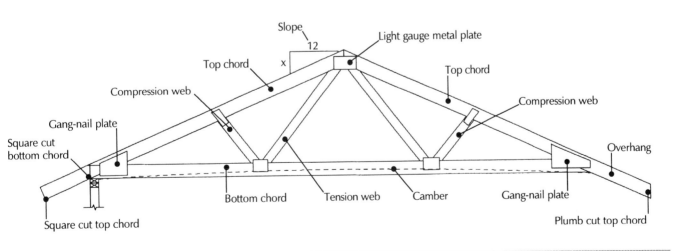

Slope
12
x

Light gauge metal plate

Top chord

Top chord

Compression web

Compression web

Gang-nail plate

Overhang

Square cut bottom chord

Bottom chord　Tension web　Camber　Gang-nail plate

Square cut top chord

Plumb cut top chord

Figure 1-28 Roof truss

You can also connect web members to the chords, as well as splices in the chord, with split rings and bolts. A split ring is a flat metal ring placed within a circular groove cut into the meeting faces of the joint. A bolt through the whole assembly holds the joint together. You can use a device called a grooving tool to cut the groove for the split ring.

Here's what you need to tell the fabricator when you order prefabricated wood trusses:

- Nominal span: length of the bottom chord
- Overhang length: horizontal distance from end of the bottom chord to the outside edge of the exterior wall
- Quantity: number of trusses required based on spacing of 24 inches oc
- Design load: top and bottom chord live and dead load and allowable stress increase
- End cut of rafter: top chord
- Roof slope: vertical rise (in inches) per 12 inches horizontal run
- Type of truss

Pitched Trusses

The most common pitched trusses include the Fink, Howe, kingpost, Pratt, queenpost, and fan truss. See Figure 1-26. Pitched trusses are the most widely used for spans up to 80 feet. This limit is based on lumber that's usually available and an average truss spacing of 15. The normal depth of a truss is $1/4$ to $1/5$ of its span.

Pratt and Howe trusses are preferred for roofs with monitors and other vertical framing projections above the truss. The vertical web members may be extended to form an integral part of the framing. Use a scissors trusses when a high ceiling clearance under the truss is required.

Flat Trusses

Slope the top chord of flat roof trusses at least $1/10$ inch per foot for drainage, although $1/4$ inch per foot is better. You should also build the truss with a camber to compensate for deflection under a load. Flat trusses can be supported by the top or bottom chord or by the intersection of web members to the bottom chord. Truss ends may be cantilevered beyond their supports.

One advantage of a flat truss is that the top of the upper chord is flush with the top of the rafters, providing a smooth surface for the roof diaphragm. Also, the bottom of the lower chord can be flush with the bottom of the ceiling joists to provide a smooth surface for the ceiling.

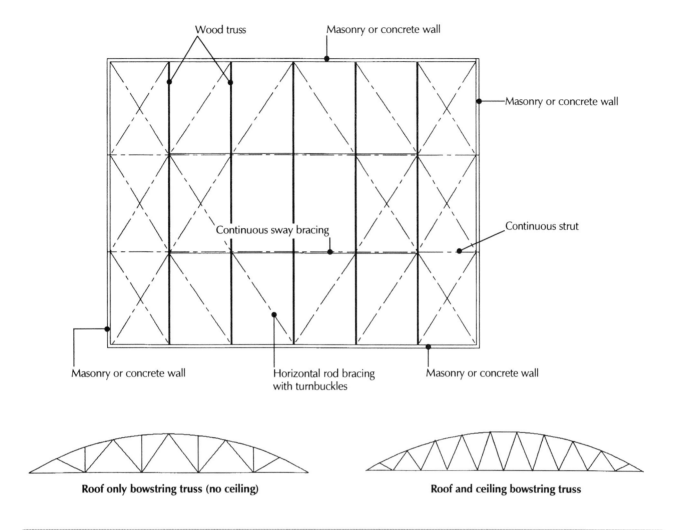

Figure 1-29 *Typical diagram of roof framing and bracing*

Bowstring Trusses

Bowstring trusses shown in Figure 1-29 provide up to 100 feet of clear span between exterior walls. Ceiling joists can be supported by the bottom chord and rafters supported by the top chord. The shape of the bowstring truss is based on the ratio of the radius to the span of the truss. The radius is determined by the angle (at the heel of the truss) from the horizontal and the distance between the top and bottom chords at the center of the truss.

The ratio of the radius and span of several common bowstring trusses are:

■ 32 degrees, radius = 0.9935 span

■ 36 degrees, radius = 0.8507 span

■ 40 degrees, radius = 0.7779 span

■ 44 degrees, radius = 0.7918 span

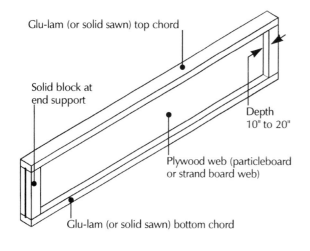

Figure 1-30 *Truss joist with solid web*

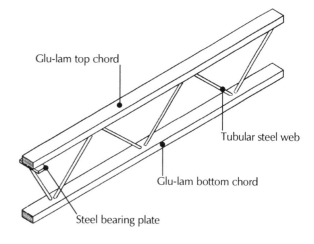

Figure 1-31 *Truss joist with open steel web*

Bracing Trusses

Brace a truss laterally to stiffen it. You need steel rod bracing to keep the top and bottom chords from bowing from a straight line and to keep the truss from tipping from a vertical plane. Use purlins, roof sheathing, and horizontal struts to brace a truss. Use plywood or diagonal board sheathing to brace the top chords, and horizontal struts or runners to brace the bottom chords.

You also need vertical sway bracing to help keep a truss from tipping or overturning. Install cross members so that one of the members will be in tension regardless of the direction of the force. Install vertical sway bracing at the locations of the horizontal runner. See Figure 1-29.

Truss Joists

Truss joists are prefabricated trusses with laminated wood top and bottom chords and a web made of ³/₈-inch structural plywood (shown in Figures 1-30 and 1-31) or steel angles, channels, or tubes. Truss joists with wood webs look like steel plate girders. The wood chords are fabricated into nominal 2 × 4s or 2 × 6s made from layers of billets glued, laminated, and pressed by a special press. The layers are coated with a waterproof phenol formaldehyde glue and then exposed to heat and pressure. The chords remain straight and won't shrink, twist, warp, or crook. The built-up section is stronger than solid-sawn wood.

Truss joists with steel webs are manufactured with parallel single taper, double taper, or curved chords. They're used mainly for roof and floor framing. These factory-built joists are made 10 to 48 inches deep and can span up to 60 feet when installed as floor joists.

Chapter 2

Wood Design Basics

Now that we've covered the composition and properties of wood structural members, we'll learn how to design the sizes of these members for a building frame.

The purpose of a structural frame is to resist outside loads. Also, the frame must be strong enough so that deflection isn't unpleasant for the building occupants. No one likes a sagging ceiling or springy floor. Before you can begin designing buildings, you have to understand the various loads.

Types of Loads

In structural design, you deal with *forces*. A force is an action that creates pressure, tension, or motion on an object. For example, the force exerted by the weight of a floor creates pressure on the beams that support the floor. Likewise, the force exerted by wind against a wall creates tension on the wind-bracing members.

A *load* is a force or combination of forces that acts vertically, horizontally, or diagonally on a structural member. A load may be concentrated at one point, such as at the base of a column. A concentrated load is designated as P in design formulas and it's usually given in pounds (lb) or kips (k). A kip is 1000 pounds. A load can also be uniformly distributed over a surface such as a floor, wall, ceiling, or roof area. This load is expressed in pounds per square foot (psf).

A horizontal member such as a beam or joist supports a floor or ceiling area (called a tributary area). This area extends laterally the same distance as the joist or beam spacing. This is shown in Figure 2-1. Since each running foot of a horizontal member (shown as beam #2 in the figure) supports a tributary area, a uniform load

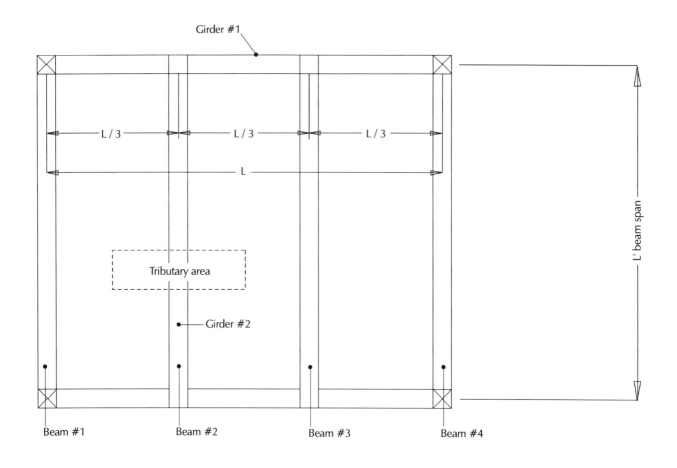

Figure 2-1 *Typical wood frame*

over a floor can be expressed as a uniform load over the member supporting the floor. In this case, the uniform load is expressed in pounds per linear foot (plf) of beam. It's usually designated as *w* in design formulas.

You can use Figure 2-2 to convert a load uniformly distributed on the beams supporting the surface from pounds per square foot to pounds per linear foot. For example, if you have a floor with a uniformly-distributed load of 50 psf supported by joists framed on 24-inch centers, the load exerted on the joists will be 100 plf. If the joist spacing is increased to 48 inches, the load exerted on the joists will be 200 plf.

Some engineers prefer to design beams using the total weight that the beam must support. For example, a 20-foot beam supporting 100 plf supports a total load of 20 × 100 = 2000 pounds. The total load supported by a beam is usually designated by *W* in the design formulas.

A structural engineer must contend with many different types of loads when he designs structural members, but he's most concerned with dead and live loads.

Dead Loads

Dead loads are loads due to the weight of all permanent building materials and fixed service equipment. For example, the dead load on a shingle roof is the combined weights of the shingles, roofing felts, decking, and rafters. Examples of fixed service equipment include HVAC units, storage tanks, and ventilators.

Figure 2-3 is a list of typical dead load values for various building components. The weight of framing members is based on spacing 16 inches on center. When the spacing is 12 inches on center, multiply these values by 1.33, and when the spacing is 24 inches on center, use two-thirds of these values.

In an office with movable partitions, figure a dead load of 20 psf throughout the floor since there is no way of knowing where these partitions will end up.

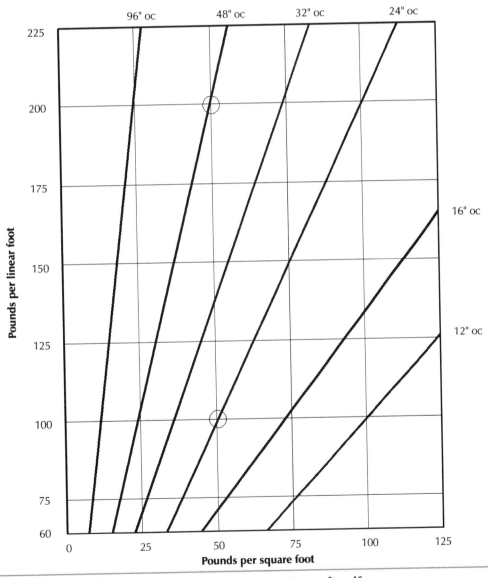

Figure 2-2 Load conversion: psf to plf

Building component (16" oc)	Dead load (psf)	Building component (16" oc)	Dead load (psf)
Built-up roofing:		**Flooring:**	
2-15 lb felts, plus 1-90 lb cap sheet	1.75	Hardwood (nominal 1")	4.0
3-15 lb felts, plus 1-90 lb cap sheet	2.2	Concrete (1" thick)	
3-ply roofing and gravel surface	5.6	Regular weight	12.0
4-ply roofing and gravel surface	6.0	Lightweight	6.0 to 10
5-ply roofing and gravel surface	6.5	Resilient flooring	1.5
		Ceramic tile ($^3/_4$" thick)	10.0
Roof and floor sheathing:		**Miscellaneous decking material:**	
$^3/_8$" plywood	1.1	Tectum (1" thick)	2.0
$^1/_2$" plywood	1.5	Insulrock (1" thick)	2.7
$^5/_8$" plywood	1.8	Poured gypsum (1" thick)	6.5
$^3/_4$" plywood	2.3	Vermiculite concrete (1" thick)	2.6
$1^3/_8$" wood sheathing	3.4	Hardwood flooring (1" nominal)	3.8
2" wood decking	4.3	Linoleum	1.5
3" wood decking	7.0	Ceramic tile ($^3/_4$" thick)	10.0
4" wood decking	9.3	**Ceilings:**	
Sprinkler system	1.0	Acoustic fiber tile	1.0
		$^1/_2$" gypsum board	2.2
		$^5/_8$" gypsum board	2.8
		Plaster (1" thick)	8.0
		Wood suspension system (with tile)	2.5
Insulation:		Metal suspension system (with tile)	10.0
Temlock (1" thick)	1.2	**Framing members:**	
Cork (1" thick)	0.7	2 x 4 at 16" oc	1.1
Gold bond (1" thick)	1.5	2 x 6 at 16" oc	1.7
Styrofoam (1" thick)	0.2	2 x 8 at 16" oc	2.2
Foamglass (1" thick)	0.8	2 x 10 at 16" oc	2.8
Rigid fiberglass (1" thick)	1.5	2 x 12 at 16" oc	3.3
Rolled rock wool (1" thick)	0.2		

Figure 2-3 Typical dead loads

Live Loads

Live loads are loads caused by the use or occupancy of the building. These include the weights of occupants, furniture, and merchandise. Commonly-used uniform and concentrated live loads are listed later in this chapter. Snow loads are also considered to be part of the live loads. We'll talk about snow loads later in this chapter.

Figure 2-4 is a suggested list to use for figuring live loads. Notice that I said *suggested*. Always consult your local building code since it will dictate the live load calculations required in your locale. Figure 2-5 is a list of the symbols and abbreviations commonly used in the design of wood structural members.

Location	Live load (psf)
Residential	40
Public rooms	100
Corridors	80
Uninhabited areas	20
Exterior balconies	60
Exit facilities	100
Retail stores	75 plus a 2000 lb concentrated load
Wholesale stores	125
Classrooms	40 plus a 1000 lb concentrated load
Light storage	100 plus a 3000 lb concentrated load
Light manufacturing	75 plus a 2000 lb concentrated load
Heavy manufacturing	125 plus a 3000 lb concentrated load
Sidewalks and driveways	250

Figure 2-4 Typical live loads

A	Area of cross section	$fc\perp$	Actual unit stress in compression perpendicular to grain
b	Breadth, or width, of a rectangular member	*Ft*	Tabulated tension stress parallel to grain
C	Coefficient, constant or factor	*Ft'*	Allowable tension stress perpendicular to grain
c	Distance from the neutral axis to the extreme fiber		
Cr	Repetitive member factor	*Fv*	Tabulated unit stress in shear parallel to grain
D	Diameter	*Fv'*	Allowable unit stress in shear perpendicular to grain
d	Depth of a rectangular member, or least dimension of a compression member		
		I	Moment of inertia
d'	Depth of beam at notch	*L*	Span length of beam, in feet
DL	Dead load	*l*	Span length of beam, in inches
E	Modulus of elasticity	*M*	Bending moment
e	Eccentricity	*P*	Total concentrated load
Fb	Tabulated unit stress for extreme fiber in bending	*r*	Radius of gyration
Fb'	Allowable unit stress for extreme fiber in bending	*Rv*	Vertical reaction
fb	Actual unit stress for extreme fiber in bending	*S*	Section modulus
		T	Total axial tension load
Fc	Tabulated unit stress in compression parallel to grain	*t*	Thickness
		V	Shear force
Fc'	Allowable unit stress in compression parallel to grain	*v*	Horizontal shear
		W	Total uniform load
fc	Actual unit stress in compression parallel to grain	*w*	Uniform load per unit of length

Note: A tabulated stress is the maximum stress specified in the building code. An allowable stress is the maximum stress permitted with limiting conditions considered. Actual unit stress is the calculated stress, or working stress. These stresses are usually given in psi.

Figure 2-5 Abbreviations and symbols used in designing structural wood members

Total Dead and Live Loads

Approximate dead and live loads on one floor of a typical wood frame multilevel condominium or apartment building are:

Dead load	Load (psf)
1^1/$_2$" lightweight concrete flooring	14.0
^{5}/$_8$" plywood subfloor	2.0
2 × 12 joists at 16 inches oc	4.0
3" insulation	0.5
Furred ceiling (over kitchen, entry)	2.0
^{5}/$_8$" gypsum board ceiling	2.5
Mechanical and electrical	1.5
Total dead load	26.5
Live load (residential)	40.0
Total dead plus live load	66.5

I've rounded up these dead loads to the next higher number to allow for the weights of miscellaneous blocking, bridging, and other minor items.

Wind Loads

Wind loads are pressure or suction forces exerted on a building due to wind. Wind causes positive pressures to the windward side of a building, and negative pressure (suction) on the leeward side. The intensity of these loads depends on the wind speed, building height, and profile. Building height is important since wind velocities usually increase with altitude. The building profile or form is important since some buildings are more aerodynamic than others. A building must be strong enough to resist wind loads from the foundation to the roof and you have to consider the effects of these loads in the design.

The negative pressure (suction) causes uplift, which does the most damage. Structural damage caused by high winds is usually the result of nails or other fasteners being forced out of the roof and wall covering. The amount of negative pressure varies at different parts of a building. There are higher uplift forces at the eaves, corners, ridges, and gable ends than at the interior areas of a roof. The uplift force increases with higher wind speeds.

There are three types of wind uplift conditions: basic, intermediate and high wind uplift. *Basic uplift* occurs in areas with wind speed under 80 mph. If you're using plywood sheathing in these areas, nail the plywood panel edges with at least 6d nails at 6 inches on center and 12 inches on center at intermediate supports.

Intermediate uplift occurs in areas with wind speeds over 80 mph. Nail plywood sheathing with at least 8d nails at 6 inch centers along panel edges, 12 inch centers along intermediate supports.

High wind uplift occurs in hurricane regions, such as the coastal areas along the Atlantic Ocean and the Gulf of Mexico. Nail plywood sheathing with at least 8d common nails at 6 inch centers at panel edges and intermediate supports. Nail plywood sheathing within 4 feet of building corners at 4 inch centers using 8d common nails. If the building is over 25 feet tall, use 8d ring-shank nails.

The basic wind pressure design formula is given in the *Uniform Building Code*. C_e and C_q tables don't appear in the IBC or IRC, but they are discussed, and there is a list of factors.

$$P = C_e \, C_q \, q_s \, I_w$$

where:
 P = Design wind pressure (psf)
 C_e = Combined height, exposure and gust factor coefficient
    (from Table 16-G, UBC and IBC)
 C_q = Pressure coefficient (from Table 16-H, UBC and IBC)
 q_s = Wind stagnation pressure at a height of 33 feet (from Table 16-F, UBC
    and IBC)
 I_w = The importance factor for building. The importance factor for a
    non-emergency building is 1.

For C_e, we'll use the value in the following table based on a 30-foot building height:

Exposure	C_e
B	0.76
C	1.23
D	1.54

The three types of exposures for wind loading are:

▮ Exposure B is a building site with adjacent buildings, forests and surface irregularities at least 20 feet high

▮ Exposure C is a building site that is flat and generally open

▮ Exposure D is a flat unobstructed area, or an area next to large bodies of water with basic wind speeds of 80 mph or greater

You can use the following table to find the values of C_q for different locations on a roof:

Location	C_q
Wind perpendicular to ridge	0.7 outward
Leeward or flat roof	0.7 outward
Windward roof less than 2 in 12	0.7 outward
2 in 12 to 9 in 12	0.9 outward and 0.3 inward
9 in 12 to 12 in 12	0.4 inward
Walls parallel to ridge	0.7 inward
Flat roofs	0.7 outward

Here are the values for q_s for various cities:

City	Basic wind speed (mph)	q_s
Los Angeles, CA	70	12.6
Amarillo, TX	80	16.4
New Orleans, LA	90	20.8
Tampa, FL	100	25.6 ←
Miami, FL	110	31.0

Now let's work an example. Assuming a 7 in 12 roof slope, find the design wind pressure for the windward side of a roof located within a B exposure area in Los Angeles where the basic wind speed is 70 mph. Assume an I_w of 1.

$$P = 0.76 \times 0.9 \times 12.6 \times 1.0$$
$$= 8.62 \text{ psf}$$

That's 8.62 psf of uplift. A 4 × 8 plywood panel (32 sf) could be subjected to a (32 x 8.62) = 276 pound uplift. If a panel is laid perpendicular to the rafters spaced 2 feet on center, every foot of rafter supports 2 square feet of panel. Uplift is 8.62 psf times 2, or 17.24 psf. Make sure you provide enough nails to resist this uplift. In Chapter 4, you'll find the withdrawal strength of nails under the heading "Connections." The withdrawal strength of 8d nails is 34 pounds, and for 10d nails it's 38 pounds.

The intensity of wind loads on a building depends on the following factors:

∎ Basic wind speed velocity (mph)

∎ Fastest-mile wind speed, measured by the National Oceanic and Atmospheric Administration (mph)

∎ Height of the exposed portion of the building above the ground (ft)

∎ Wind direction

Figure No. 16-1 in the 1997 *Uniform Building Code* (our Figure 2-6) is a map of United States showing the minimum basic wind speeds in miles per hour. Similar maps are in the 1993 *BOCA* and the 1994 *Standard Building Code*. But tornadoes are generally beyond the scope of the building code. In 1994, a tornado hit Lancaster, Texas with wind velocity in excess of 200 mph. It destroyed 80 percent of the town. Damages included:

∎ Wind lifted built-up roofing and roof decks, causing trusses to collapse.

∎ Wind-borne debris broke windows and tore holes in walls.

∎ Mature trees were uprooted and completely debarked.

∎ Brick masonry walls collapsed.

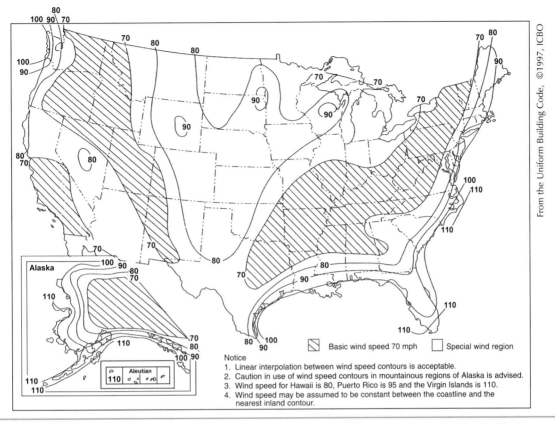

Figure 2-6 *Minimum basic wind speeds in miles per hour (× 1.61 for km/h)*

The area was assigned a basic wind speed of 70 mph by the building code, so the design and construction of the buildings within this area didn't anticipate the ferocity of winds associated with a tornado. Conventional construction isn't designed to withstand major hurricanes and tornadoes. It would make houses too expensive for most people to afford.

Snow Loads

There are many areas in the United States where snow load is a significant part of roof design. See Figure 2-7. Snow loads in the northern parts of the United States can be as great as 70 psf. Your building code may include a map of your area showing ground snow loads. Figures A-16-1 through A-16-3 in the UBC (our Figure 2-8) and Figure R301.2(5) in the IRC, show the ground snow load (P_g) for a 50-year mean interval for different areas in the U.S.

You'll need to consider the following factors when you figure snow loads:

▌ Building location and exposure

▌ Roof slope

▌ Importance of building (emergency versus non-emergency building)

▌ Density of snow, which varies with the amount of water mixed with the snow

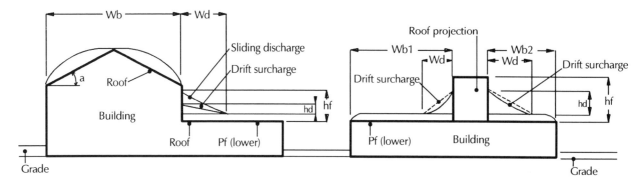

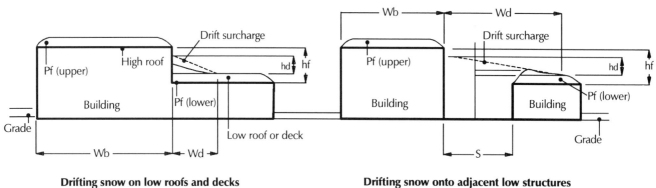

Figure 2-7 Snow loads

- Depth of snow

- Drift conditions

- Adjacent structures, such as a high roof next to a low roof on the same building

Use this formula to find the snow load *(Pf)* on flat or low-slope roofs inclined less than 30 degrees:

$$P_f = C_e \times I \times P_g$$

where:

P_f = Minimum roof snow load (psf)

C_e = Snow exposure factor (0.6 in open terrain extending a half mile or more from the building, 0.9 in densely forested or sheltered area, and 0.7 in all other conditions)

I = Importance factor

P_g = Base ground snow load which varies from 0 to 70 psf (Figure 2-8)

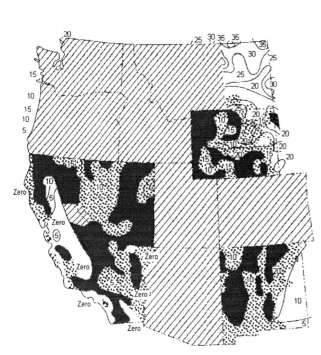

A UBC Figure A-16-1, Western U.S.

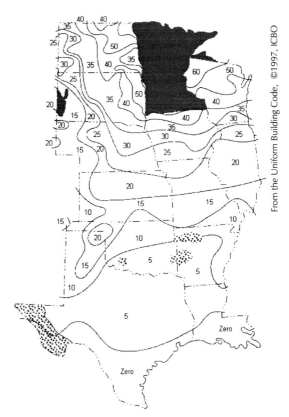

B UBC Figure A-16-2, Central U.S.

 In these areas, extreme local variations in snow loads preclude mapping at this scale; ground snow load, P_g, shall be established by the building official.

 The zoned value is not appropriate for certain geographic settings, such as high country; in these areas, ground snow loads, P_g, shall be established by the building official.

 In these areas, ground snow load, P_g, shall be established by the building official.

 Those areas shown as 0 psf, 5 psf and 10 psf (0 N/m^2, 239 N/m^2 and 479 N/m^2) are for information only.

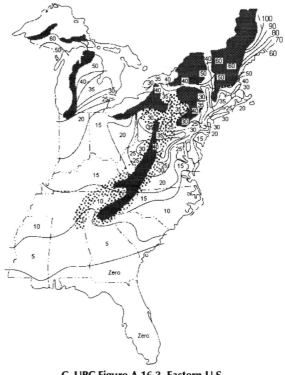

C UBC Figure A-16-3, Eastern U.S.

Figure 2-8 *Ground snow load,* P_g*, for 50-year mean recurrence interval for the United States, pounds per square foot (x 47.8 for N/m²)*

For roof slopes inclined over 30 degrees, use this formula to figure the snow load (P_s):

$$P_s = C_s \times P_f$$

where:

P_s = Sloped roof snow factor (psf)

C_s = Slope reduction factor: $1 - \dfrac{(a-30)}{40}$

a = Roof slope (degrees)

P_f = Flat roof snow load (psf)

As an example, assume a house with a 40-degree roof slope is located in open terrain in northern Michigan with an exposure factor of 0.60 and a ground snow factor of 70 psf. Then the slope factor is:

$$
\begin{aligned}
C_s &= 1 - \frac{(a-30)}{40} \\
&= 1 - \frac{(40-30)}{40} \\
&= 1 - 0.25 \\
&= 0.75
\end{aligned}
$$

Assuming an importance factor of 1, the flat roof snow load is:

$$
\begin{aligned}
P_f &= C_e \times I \times P_g \\
&= 0.6 \times 1 \times 70 \text{ psf} \\
&= 42 \text{ psf}
\end{aligned}
$$

So the sloped roof snow factor is:

$$
\begin{aligned}
P_s &= C_s \times P_f \\
&= 0.75 \times 42 \text{ psf} \\
&= 31.5 \text{ psf}
\end{aligned}
$$

Look at Figure 2-7 to see how particularly nasty things can get when you have to deal with various combinations of roofs and adjacent buildings. The variables shown in the illustration are:

S = Horizontal separation between adjacent buildings (ft)

h_b = Height of balance snow load on lower roof or deck (ft)

h_r = Difference in height between upper and lower deck (ft)

h_d = Maximum height of drift surcharge (ft)

W_b = Horizontal dimension of upper roof, not less than 25 feet

W_d = Width of snow drift (ft)

D = Density of snow (pcf)

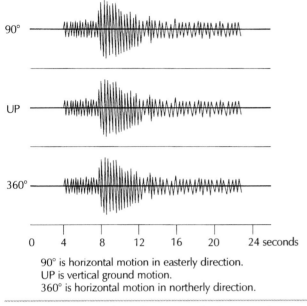

90° is horizontal motion in easterly direction.
UP is vertical ground motion.
360° is horizontal motion in northerly direction.

Figure 2-9 *Three-dimensional seismograph*

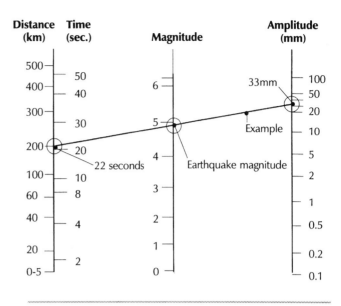

Figure 2-10 *Earthquake magnitude chart*

Seismic Loads

Seismic loads are horizontal and vertical forces on a building caused by sudden earth movement under the foundation. The magnitude of an earthquake is an index of the energy released at its epicenter. This energy can be recorded on a seismograph (Figure 2-9) and measured on the Richter scale. Higher Richter numbers mean longer and stronger shaking. The magnitude is determined by the relationship between the amplitude of ground movement at the instrument's location and the distance to the epicenter. This is shown on Figure 2-10. The amplitude is measured from the seismograph plot as shown on Figure 2-11.

Ground shaking intensity and damage caused by an earthquake are also measured on the Mercalli scale or the Rossi-Forel scale. These scales describe the effects of the earthquake on people and buildings at a specific location. For example, the quake may be described as "a slow rolling motion," or "a violent shaking."

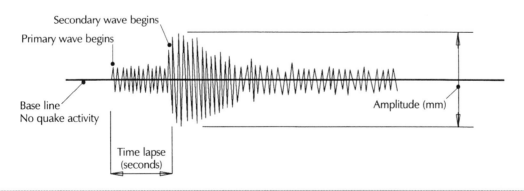

Figure 2-11 *Seismograph plot*

Types of Earthquakes

The following list describes the physical effects of twelve types of earthquakes defined by the Mercalli intensity scale:

Type I: Felt by very few, under especially favorable circumstances. (Equal to Type I of the Rossi-Forel scale)

Type II: Felt only by a few persons at rest, especially on upper floors of buildings. Delicately suspended objects may swing. (Types I and II, Rossi-Forel scale)

Type III: Felt quite noticeably indoors, especially on upper floors of buildings but many people don't recognize it as an earthquake. Standing automobiles may rock slightly. Vibration is like a passing truck. (Type III, Rossi-Forel scale)

Type IV: Felt indoors during the day by many; felt outdoors by few. Some awakened at night. Dishes, windows and doors disturbed; walls making creaking sound. Sensation like heavy truck striking building. Standing cars rock noticeably. (Type IV and V, Rossi-Forel scale)

Type V: Felt by nearly everyone; many awakened. Some dishes, windows, etc. broken; a few instances of cracked plaster; unstable objects overturned. Disturbances of trees, poles and other tall objects sometimes noticed. Pendulum clocks may stop (Type V and VI, Rossi-Forel scale)

Type VI: Felt by all; many frightened and run outdoors. Some heavy furniture moved; a few instances of fallen plaster or damaged chimneys. Damage slight. (Type VI and VII, Rossi-Forel scale)

Type VII: Everybody runs outdoors. Damage negligible in buildings of good design and construction; slight to moderate in well-built ordinary structures; considerable in poorly-built or badly-designed structures; some chimneys broken. Noticed by persons driving automobiles. (Type VIII, Rossi-Forel scale)

Type VIII: Damage slight in specially-designed structures; considerable in substantial buildings, with partial collapse; great in poorly-built structures. Panel walls thrown out of frame structures. Fall of chimneys, factory stacks, columns, monuments, walls. Heavy furniture overturned. Changes in well water. Disturbance felt by persons driving automobiles (Type VIII to IX, Rossi-Forel scale)

Type IX: Damage considerable in specially-designed structures; well-designed frame structures thrown out of plumb; great in substantial buildings, with partial collapse. Buildings shifted off foundations. Ground cracked conspicuously. Underground pipes broken (Type IX+, Rossi-Forel scale)

Type X: Some well-built wood structures destroyed; most masonry and frame structures destroyed, including foundations; ground badly cracked. Rails bent. Landslides considerable from river banks and steep slopes. Shifted sand and mud. Water splashed (slopped) over banks. (Type X, Rossi-Forel scale)

Type XI: Few, if any, masonry structures remain standing. Bridges destroyed. Broad fissures in ground. Underground pipelines completely out of service. Earth slumps and land slips in soft ground. Rails bent greatly.

Type XII: Damage total. Waves seen on ground surface. Lines of sight and level distorted. Objects thrown upward into the air.

The most common way of describing earthquake intensity is the Richter Magnitude Scale. A severe earthquake on this scale is one with a magnitude of 6.1 to 8.2 or higher. The seismographic plot in Figure 2-11 shows three types of waves in sequence. Prior to an earthquake the baseline is a steady, even line. The first set of waves is the primary wave. The second set is the secondary wave which begins with a high amplitude and tapers down. This is followed by the long waves.

The elapsed time between the beginning of the primary wave and the beginning of the secondary wave indicates the distance of the epicenter from the instrument. For example, an elapsed time of 22 seconds is equal to about 200 kilometers. The maximum amplitude of the secondary wave is measured on the chart in millimeters. Seismographers use these two figures to calculate the magnitude of an earthquake.

A straight line drawn between the distance and the amplitude marks shows the magnitude. Figure 2-10 is a magnitude chart showing an elapsed time of 22 seconds and an amplitude of 33 mm. That produces a magnitude of 4.9 on the Richter Scale.

Designing for Seismic Forces

Structural design for seismic forces depends on the building's location, framing system, building weight, configuration, and height. Seismic design also depends on the soil condition and the importance of the building to the public safety during a catastrophe. All of these factors are used in a formula that gives us the maximum horizontal force at the base of a building. This force is called base shear (V) and the formula for base shear is:

$$V = \frac{Z \times I \times C \times W}{R_w}$$

where:

V = Total lateral shear force at the building's base

Z = Seismic coefficient for earthquake probability zone (UBC Figure 16-2, our Figure 2-12), and as discussed in the IBC and IRC.

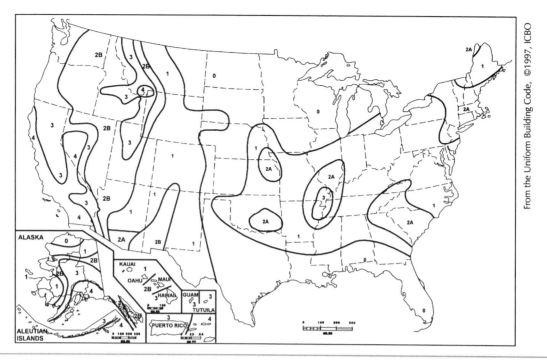

Figure 2-12 *Seismic zone map of the United States*

I = Importance factor (1 for non-emergency building and 1.5 for emergency buildings)

C = Numerical coefficient for natural period of vibration of building, not to exceed 2.75

W = Total dead load of building

R_w = Numerical coefficient for the structural system (8 for light-framed walls with shear panels)

Figure 2-13 is a simplified description of what happens to a three-story building during one cycle of earth movement in an earthquake. In reality, the ground rocks, twists, heaves and subsides, changing direction and speed simultaneously. The building tries to follow these violent ground movements within the range of its flexibility. A perfectly elastic building would return to its original shape, but brittle components of an actual building will rack and break. With properly-designed and well-built connections, the building should hold together at least long enough for the occupants to leave. This is the basis of "life safety" design.

Seismic loads occur when a building is moved suddenly during an earthquake, causing the ground and foundation to pitch and roll. The building is driven horizontally as well as vertically, resulting in acceleration. Seismic acceleration is commonly expressed as a decimal part of acceleration caused by gravity *(g)* which is 32.2 feet per second per second. So a structural engineer might design building parts to withstand a seismic force of 0.13g, 0.33g, 0.50g or more. In other words, different members of a building might be designed to withstand horizontal forces equal to 13, 33 or 50 percent of the member's weight.

Seismic design has changed from relatively simple formulas in the late 1930s to complex formulas used in the latest building codes. As more earthquakes occurred and the data was analyzed, the formulas were revised several times. The seismic force formula is:

$$V = \frac{Z \times I \times C}{R_w} W$$

As an example, find the lateral shear force for a non-emergency wood frame building in Los Angeles (Zone 4).

$$V = \frac{0.40 \times 1 \times 2.75}{8} W$$
$$= 0.1375 \times W$$

Therefore, the total base shear of the building is about 13.75 percent of the total dead weight of the building.

Lessons from the Northridge Earthquake

The 1994 earthquake in Northridge, California was a wake-up call for poor wood-frame construction. An extensive investigation was conducted by the Wood Frame Subcommittee of the Department of Building and Safety and the Structural Engineers Association of Southern California (SEASC). They found that portland cement plaster (stucco) and drywall shear walls performed poorly in the many multistory apartments and condominiums during the Northridge earthquake. Here are some of the types of framing failures that resulted from poor or inadequate framing:

▌ Some narrow walls rotated and failed because wood diagonal corner let-in bracing was used instead of plywood sheathing.

▌ Some walls bowed out because 1 × 4 diagonal braces weren't let-in or adequately nailed to the studs.

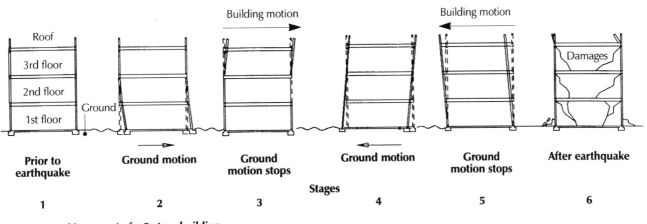

Movement of a 3-story building
1. Prior to earthquake
2. Foundation moves right. Building starts to follow.
3. Foundation stops. Upper part of building continues right.
4. Foundation moves left. Building stops moving right.
5. Foundation stops. Building moves left.
6. Earthquake stops. Building damaged.

Figure 2-13 Building sway

- Unsheathed walls rotated because they lacked let-in braces.

- Exterior lath and plaster separated from studs because the lath wasn't nailed to the studs with enough nails, or the nails used were too short.

- Brick chimneys fell down because they weren't adequately anchored to the ceiling joists.

- Interior gypsum board shear walls failed because of inadequate nailing and edge blocking.

- Connections between the wood roof deck and masonry walls failed because of inadequate anchorage and nailing into the ledger.

- Rotational stresses occurred in asymmetrical buildings, causing portions of the building to separate due to rotation around shear walls of differing stiffness.

- Interior gypsum board walls acted like unintended shear walls and failed. Don't install plywood shear walls in line with gypsum board shear walls.

- Plywood shear walls failed because of inadequate or undersized nailing, or because nails were driven too close to the panel edge.

- Major structural failures occurred because there was a lack of connectors for continuous tension ties across intersecting framing members for roof and floor diaphragms.

- Heavy HVAC equipment on roofs shifted and over-stressed and stretched their bolted connectors due to vertical acceleration.

After evaluating the damage, the Los Angeles City Building and Safety Department made the following changes in earthquake design of shear walls in wood-frame buildings:

- Reduce the shear value of portland cement plaster from 180 plf to 90 plf.

- Reduce the shear value of all gypsum sheathing boards to 30 plf, although the 1977 UBC and 2000 IBC allow 100 to 175 plf for gypsum boards of various thicknesses and nailing.

- Limit the height to depth ratio (h/d) of gypsum wallboard shear walls from 2:1 to 1:1.

- Reduce the h/d ratio of plywood shear walls from $3^{1}/_{2}$: 1 to 2:1.

- Prohibit use of portland cement plaster, gypsum board sheathing, or gypsum wallboard shear walls to carry shear loads on the ground floor of multi-level buildings. Allow only the use of plywood.

- Use only common nails.

- Use 3x members with a shear value of at least 300 plf for sill or sole plates and studs supporting panel edges.

- Increase the distance from the panel edges to the nails from $^{3}/_{8}$ to $^{1}/_{2}$ inch.

- Sheath cripple walls between the foundation and first-floor framing with plywood.

- Provide adequate anchor bolts for all mud sills.

- Provide adequate hold-downs for all shear walls.

So what's adequate for anchor bolts and hold-downs? Use at least $1/2$-inch-diameter anchor bolts or bolts for hold-downs embedded at least 7 inches into the concrete. Install a minimum of two anchor bolts per piece of mud sill with one bolt located within 12 inches of the end of each piece.

Impact Loads

Impact loads are sudden applications of forces caused by mechanical devices such as elevators or cranes. A crane can cause both horizontal and vertical impact loads on a structure.

Special Loads

Special loads are temporary loads like construction walkways, canopies, and handrails. Movable partitions are also considered special loads.

Miscellaneous Loads

Miscellaneous loads are weights of minor service equipment such as fire sprinkler piping, lighting fixtures, and HVAC ducts suspended from a roof, ceiling or floor. The total weight of any service equipment may be spread over the area of its mounting platform.

Combination Loads

Combination loads are simultaneous loads acting on a building frame and may consist of:

- Dead and live floor load plus dead and live roof load

- Dead and live floor load plus wind load (or seismic load)

- Dead and live floor load plus wind load plus one-half snow load

- Dead and live floor load plus snow load plus seismic load

Eccentric Loads

Posts are most effective when the load on them is applied concentrically. When the load is off the center axis, the compression member is eccentrically loaded. This results in a combined compression and bending load that reduces the carrying capacity of the post. The formula to calculate the combined stress on a post is:

$$f = \frac{P}{A} \pm \frac{M}{S}$$
$$= \frac{P}{A} \pm \frac{Pe}{S}$$

where:

f = Combined stress on post (psi)
P = Axial load (lbs)
A = Cross-sectional area of post (in^2)
M = Bending moment = P × e (inch-lbs)
e = Eccentricity (in)
S = Section modulus of post (in^3)

Stresses in Wood Members

To select a proper structural member, you must know the kinds of stresses the member will be subjected to. Stress is a force applied over a unit area, usually given as pounds per square inch, or psi. If we deal with thousands of pounds, stress is given in kips per square inch, or ksi. Figure 2-14 illustrates some of kinds of stresses.

The important types of stresses you must consider while designing structural members include bending stress, shear stress, compression stress and tensile stress. Figure 2-5 includes the symbols usually used for these stresses.

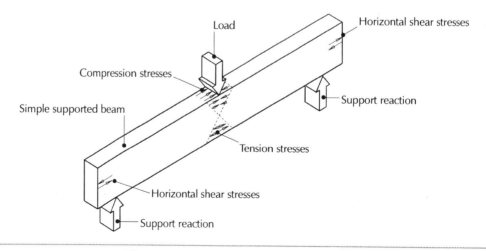

Figure 2-14 *Simple beam stresses*

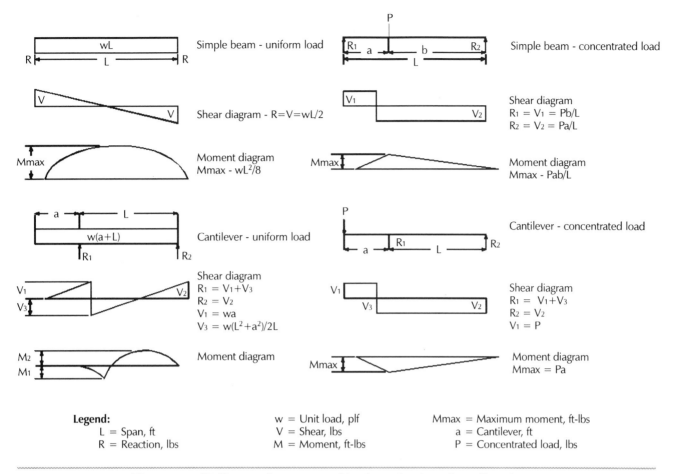

Legend:

L = Span, ft	w = Unit load, plf	Mmax = Maximum moment, ft-lbs
R = Reaction, lbs	V = Shear, lbs	a = Cantilever, ft
	M = Moment, ft-lbs	P = Concentrated load, lbs

Figure 2-15 Uniform and concentrated loads on simple and cantilever beams

Bending Stress

A beam bends when it's loaded. The maximum bending stress in a uniformly loaded beam supported at each end is at the center of the beam. Look at the moment diagram for the simple beam shown in Figure 2-15. You'll see that the bending moment is greatest at the center of the beam's span, which is what you would expect. Figure 2-15 also shows simple and cantilever beams with uniform and concentrated loads.

Bending results in compression in the upper half of the beam and tension in the lower half. The stress is greatest on the wood fibers that are the farthest from the center of the beam, so the top fibers of the beam have the maximum compressive stress and the bottom fibers have the greatest tensile stress. There is zero stress at the neutral axis at the center of the cross section of the beam. Since the edges of the beam are the weak links, bending (or flexural) stress used for design purposes is allowable bending design value and is designated as F_b'.

You have to contend with two sets of values for F_b, depending on whether a beam is used as a single or repetitive member. Single-member use means that the beam carries a load with no assistance from adjacent structural members. Repetitive-member use applies to members such as joists and rafters that are spaced no more than 24 inches on center and joined together by a subfloor, roof deck, or other load-distributing area. For example, F_b for Douglas fir-south select structural is equal to

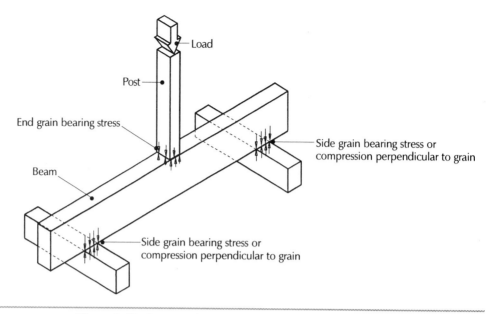

Figure 2-16 Compression stresses

1,300 psi for normal load duration and dry service condition. But you have to adjust this by the repetitive member factor $C_r = 1.15$ when it's used for joists, rafters, decking, and similar building elements. Then allowable value for F_b is $1,300 \times 1.15$ or $1,495$ psi.

Shearing Stress

Shearing stress results from two opposite but parallel forces that cause adjoining surfaces of a member to slide past one another, or to tear. *Vertical shear* is the stress that resists a beam's tendency to shear off (or split) vertically on each side of a support. Look at the vertical shear diagram for the simple beam in Figure 2-15. You can see that the vertical shear forces are greatest at the supports, and zero at the center of the beam.

Horizontal shear is the stress that resists the tendency of the wood fibers of the beam from shearing by splitting and sliding past each other horizontally. Since shear failures are due to horizontal shear, it's the weak link and is designated by F_v.

Compressive Stress

Compressive stress is the stress that results from a force that tries to crush a member. Compressive stress can be parallel or perpendicular to the grain. An example of compressive stress parallel to grain is a column or post loaded from above and it's designated by F_c. The top half of a loaded beam is an example of compression parallel to grain.

Common framing members that carry compression loads are studs, columns, posts, and struts. Studs carry the load from ceiling joists and rafters. Columns and posts support beams or lintels and are usually made of solid-sawn 4 × 4 or 6 × 6 lumber. Posts can also be built up with two 2 × 4s or three 2 × 6s. Figure 2-16 shows allowable compressive stresses for posts and beams. There's more on the design of beams later in the chapter.

Wood piles that support building foundations are examples of structural members subjected to compression. They can be supported at their tips by bedrock or by sandy soil through friction. You should load piles concentrically to prevent bending and buckling. That is, design the foundation so that the load is applied to the center of the pile and the force acts along the vertical axis of the pile. See Figure 2-17 for typical wood piling. You should use pressure-treated piles unless the piles will be continuously under the lowest ground water level. Wood won't rot as long is it's always submerged. The allowable unit stress in round wood piles shouldn't exceed:

Species	F_c' psi	F_b' psi	F_v' psi	F_c' psi
Douglas fir	1,250	2,450	115	230
Southern pine	1,200	2,400	110	250

The modulus of elasticity (E) for both species is approximately 1,500,000 psi.

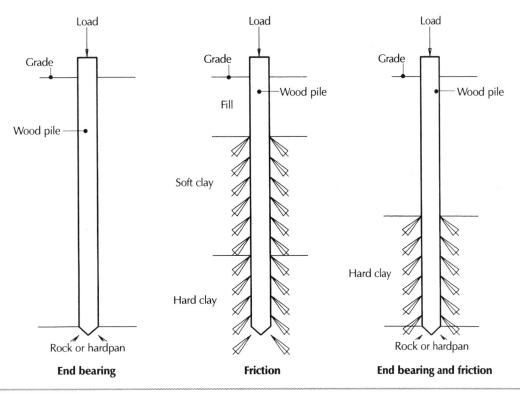

Figure 2-17 *Load transfer from wood piles to soil*

Tensile Stress

Tensile stress results from a force that tends to elongate a member. Wood is relatively weak when tension is perpendicular to its grain, so it's poor engineering practice to design or install a member with tension in this direction. That's why tension parallel to the grain is the only property considered in wood design and is designated by F_t.

Framing members subjected to tensile stress include rafter ties, wall bracing, drag struts, and the lower chords of trusses. Rafter ties resist the outward thrust of the rafters. Wall bracing resists wind or earthquake loads by tension (when forces stretch the wall) and by compression (when forces squeeze the wall).

Factor of Safety

The ratio between the ultimate unit stress (the stress that would actually cause failure of the member) and the allowable unit stress is the factor of safety. For wood, the factor of safety is about 1 in 8 or 10. In other words, we use about $1/8$ to $1/10$ of the ultimate strength of wood for design. A safety factor is necessary in wood design because wood construction is so variable. For example, strength is lost due to defects, size and location of knots, slope of grain, lumber density, and moisture content. Careless bolting or nailing also reduces the strength of a wood member.

Figure 2-18 shows various ways that a beam can fail under excessive loading.

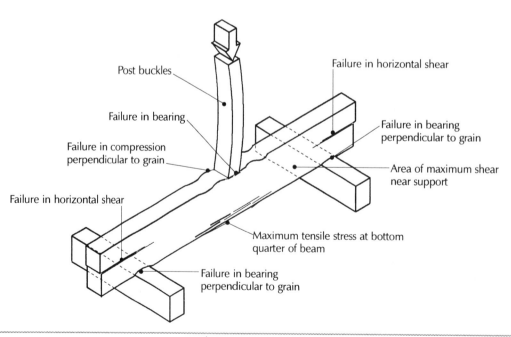

Figure 2-18 Stress failures

Duration and type of load	Adjustment factor (DOL x F_b)	Adjusted F_b, psi
Permanent (dead load)	0.9 x 1500	1350
10 years (floor load, occupancy)	1.0 x 1500	1500
2 months (snow load)	1.15 x 1500	1725
1 week (construction load)	1.25 x 1500	1875
10 minutes (wind/earthquake load)	1.33 x 1500	2000
Impact (impact load)	2.0 x 1500	3000
Note: F_b = 1500 (for duration of load)		

Figure 2-19 Adjustment of allowable stress

Load Factors

Normal loading is a design load that stresses a member to the full allowable stress tabulated in the building code. Wood is a resilient material that can support a heavier load over a short period of time than it can over a longer period. Therefore, allowable stresses in wood can be adjusted by using a load factor for the duration of the loading condition. Use the adjustment factors shown in Figure 2-19 to adjust the allowable stresses under different loading conditions.

As an example, if the allowable bending stress (F_b') of a wood member is 1,500 psi under normal loading conditions, you can assume that the allowable bending stress can be increased to 1,500 × 1.25, or 1,875 psi for temporary construction loads.

Engineering Properties of Wood Members

We need to discuss a few physical properties of wood before we actually design beams and columns. These properties include the strength of a member, its moment of inertia, bending moment, section modulus, and modulus of elasticity. These might sound intimidating, but don't be discouraged. I'll simplify these terms so that you can work the design problems like a pro after a bit of practice.

For any job, the size of framing member you need depends on its strength (species and grade), the total load it supports, its center-to-center spacing, and its span.

The cross-sectional area *(A)* of a structural member is the total amount of material available to resist axial forces imposed on it. Keep in mind that the cross-sectional area is figured from the member's *dressed* dimensions. The member may be a solid-sawn shape or built up out of several smaller boards.

Many physical properties of the member are based on its net cross-sectional area and the lengths of its two principal neutral axes. The vertical neutral axis is called the y-axis and the horizontal axis is called the x-axis.

Moment of Inertia

Moment of inertia (I) of a member is a measure of its ability to resist bending forces. The larger the moment of inertia, the stronger the member is at resisting bending. Figure 2-20 shows the moment of inertia for common sizes of structural members.

Moment of inertia is expressed in inches to the fourth power or *(in⁴)*. It's based on the neutral axis passing through the center of the member separating the compressive and tensile stresses. The formula for *I* through the neutral axis of a member with a rectangular end area is:

$$I = \frac{bd^3}{12}$$

where:
b = dressed width of member (inches)
d = dressed depth of member (inches)

Notice that the moment of inertia depends *only* on the size of a member. For example, a 2 × 4 of any species has a moment of inertia of:

$$I = \frac{1.5 \times 3.5^3}{12}$$
$$= 5.359$$

Bending Moment

Bending moment is the measure of the tendency of a beam to bend or flex under a load. The bending moment *(M)* of a simple beam with a uniformly distributed load is:

$$M = \frac{W \times L}{8}$$

where:
W = total uniformly distributed load (lbs)
L = span

Bending moment is expressed in foot-pounds when *L* is measured in feet, or in inch-pounds when *L* is measured in inches. Bending moment can also be determined by:

$$M = \frac{w \times L^2}{8}$$

Nominal size (in)	Moment of inertia (in⁴)	Section modulus (in³)
2 x 4	5.36	3.06
2 x 6	20.80	7.56
2 x 8	47.64	13.14
2 x 10	98.93	21.39
2 x 12	177.98	31.64
2 x 14	290.76	43.89
4 x 4	12.51	7.15
4 x 6	48.52	17.65
4 x 8	111.15	30.66
4 x 10	230.84	49.91
4 x 12	415.28	73.83
4 x 14	678.46	102.41
4 x 16	1034.42	135.66

Figure 2-20 Moment of inertia and section modulus of members

where:

w = load (plf)

L = span (ft)

Section Modulus

The section modulus *(S)* of a member indicates its bending strength. You determine the section modulus by dividing the moment of inertia *(I)* by the distance from the neutral to the outer edge *(c)* of the section. See Figure 2-21. Keep in mind that *c* is the distance from the center of the board to the edge measured parallel to the long edge of the board. Use Figure 2-20 to find the section modulus of common structural members. The section modulus depends only on the size of a board. For example, a 2 × 4 of any species has a section modulus of 3.06.

You can use either *I* or *S* to find the actual bending stress *(fb)* in a beam when the bending moment *(M)* is known. Here are two formulas using moment of inertia or section modulus to find bending stress *(fb)*:

$$f_b = \frac{M \times c}{I} \ \text{ or } \ f_b = \frac{M}{S}$$

Modulus of Elasticity

The modulus of elasticity is a measure of the performance of a beam with respect to deformation. It varies by species and grade. Testers find the modulus of elasticity of a material by placing it in tension (or compression) by applying a known force (unit stress). Then they measure the unit deformation (elongation or contraction)

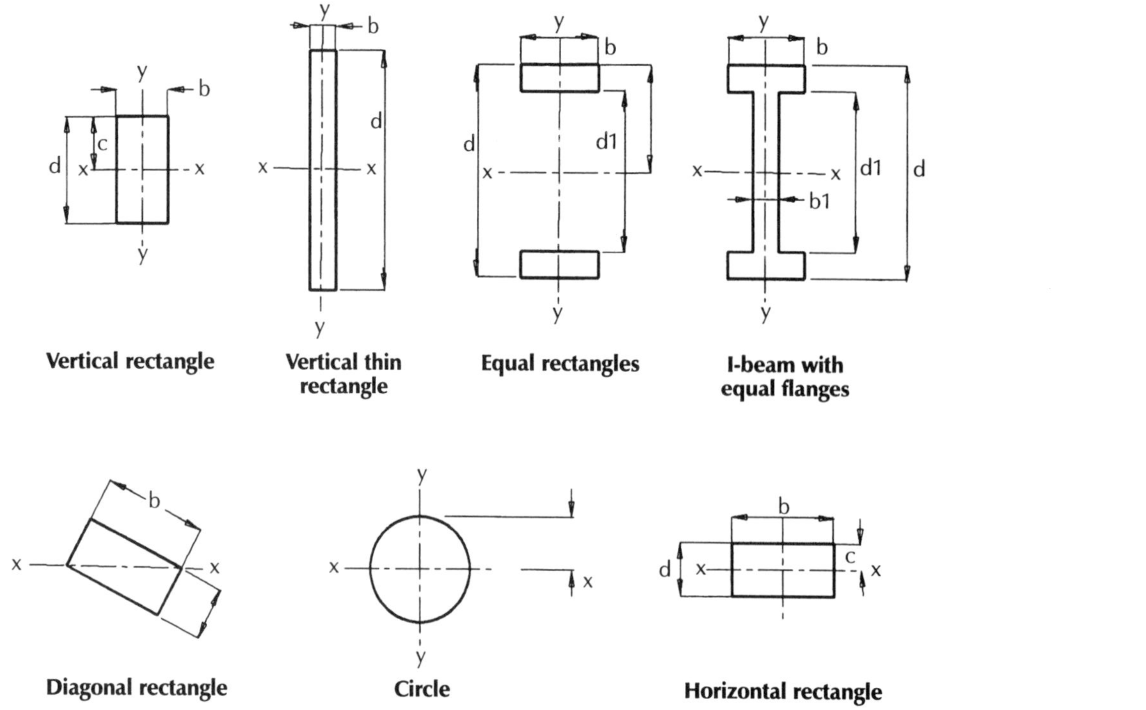

Figure 2-21 *Properties of sections*

caused by that amount of force. The modulus of elasticity of the material is the ratio between the unit stress applied and the unit deformation suffered. The modulus of elasticity for various wood species is shown in Figure 2-22. But keep in mind that the modulus of elasticity depends on the grade of the lumber.

Designing Girders, Beams and Joists

Here are the basic rules for designing girders, beams, and joists:

- Framing members must be strong enough to support the given loads.
- The member must be strong enough so that deflection isn't excessive.
- The depth of the member must be within the limits set by the architectural design.
- The design should include board sizes that are easily purchased locally.

The engineer must also take into consideration whether a beam supports a uniformly-distributed load or a concentrated load. In this chapter, we'll study a uniformly-distributed load to keep the calculations simple. For the same reason, when we discuss loads on columns, we'll talk about concentric (or axial) loading as

Common names of species	Specific gravity	Modulus of elasticity (psi)
Alder, red	.37	1,170,000
	.41	1,380,000
Aspen, bigtooth	.36	1,120,000
	.39	1,430,000
Basswood, American	.32	1,040,000
	.37	1,460,000
Beech, American	.56	1,380,000
	.64	1,720,000
Birch, paper	.48	1,170,000
	.55	1,590,000
Oak, red	.68	2,280,000
Walnut, black	.51	1,420,000
Douglas fir, coast	.45	1,560,000
Fir, California red	.36	1,170,000
Larch, western	.48	960,000
	.52	1,870,000
Pine, eastern red	.34	1,360,000
Pine, ponderosa	.38	1,000,000
Spruce, black	.38	1,060,000
	.40	1,530,000
Tamarack	.49	1,240,000
	.53	1,640,000
Southern pine	.48	1,500,000
	.52	2,000,000

Figure 2-22 Modulus of elasticity of wood species

compared to eccentric loading. Concentric loading simply means that the load is considered to go down through the center of the cross section of the column. We'll also limit our discussion by considering simple beams which are supported only on both ends. To see how nasty things can get when you consider a more complicated beam, look at Figure 2-15 which shows typical combinations of loads on simple and cantilever beams. A simple beam is shown in the upper left portion of the drawing.

Designing a Uniformly-Loaded Simple Beam

Use the following steps to design a uniformly-loaded simple beam:

▌ Find the combined live and dead load per square foot.

▌ Multiply this load by the spacing of the supports to get the uniform load, w, in pounds per linear foot (plf).

■ Multiply w by the span *(L)* and get the total load *(W)*. Maximum bending moment, M, is equal to the total load, W, times the span divided by 8, or:

$$M = \frac{W \times L}{8}$$

Bending moment is expressed in foot-pounds (ft-lb). Be sure you always use the same units in the formula, whether feet or inches, pounds or kips.

There are three basic tests to perform when you design a uniformly-loaded wood beam:

■ Check for maximum bending stress.

■ Check for maximum horizontal shear.

■ Check for maximum deflection.

In other words, check for the weak link.

Go through the following steps to find the bending stress in a uniformly-loaded beam as shown in Figure 2-15:

1) Find the span *(L)* where L is distance between center of supports.

2) Determine the uniform load *(w)* where w is the weight in pounds per linear foot of beam.

3) Calculate the maximum bending moment *(M)* where $M = \frac{w \times L^2}{8}$

4) Select a trial beam size and find the depth *(d)* and breadth *(b)*.

5) Find the moment of inertia *(I)* of the beam where $I = \frac{bd^3}{12}$

6) Determine the section modulus of the beam *(S)* where $S = \frac{I}{c}$ and $c = \frac{d}{2}$

7) Calculate the actual bending stress *(fb)* where $f_b = \frac{M}{S}$

8) If f_b is greater than the allowable stress *(Fb')*, select a larger member and try steps 1 through 7 again.

To check horizontal shear:

1) Find maximum shear *(V)* where: $V = R = \frac{w \times L}{2}$

2) Calculate the horizontal shear *(v)* where: $v = \frac{3V}{2bd}$

3) If the shear stress is higher than allowable stress, select a larger member and try again.

If there's a notch in the beam near a support, check the stress at the notch with the following formula:

$$F_v = \frac{3V}{2bd'} \times \frac{d'}{d}$$

where:

F_v = shear stress at notch

b = breadth of beam

d = depth of beam

d' = depth of beam at notch

To check deflection:

1) Determine the modulus of elasticity *(E)* of the member from Figure 2-20.

2) Determine the total load on the member *(W)* where: W = w × L

3) Calculate deflection *(D)* where: $D = \frac{5 \times W \times L^3}{384 \times E \times I}$

4) If *D* is greater than $^1/_{360}$ of *L*, select a deeper member. Otherwise, the member is acceptable.

Example

As an example, let's size floor beams spaced 4 feet on center and spanning 20 feet. These beams carry 2 × 6 wood decking and a live load of 40 psf. According to Figure 2-3, the decking weighs 4.3 psf. So the uniform load on each beam is 4 × (40 + 4.3) or 177.2 plf. Round that up to 200 plf to account for miscellaneous items and the weight of the beam.

The maximum bending moment on each beam is:

$$M = \frac{w \times L^2}{8}$$
$$= \frac{200 \times 400}{8}$$
$$= 10,000 \text{ ft-lbs or } 120,000 \text{ in-lbs}$$

The maximum shear is:

$$V = R = \frac{w \times L}{2}$$
$$= \frac{200 \times 20}{2}$$
$$= 2,000 \text{ lbs}$$

Now let's try a 4 × 10 beam for size. According to Figure 2-20, the section modulus of a 4 × 10 is 49.91 in³. Therefore, the maximum bending stress is:

$$f_b = \frac{M}{S}$$
$$= \frac{120,000}{49.91}$$
$$= 2,404 \text{ psi}$$

The bending stress is too high. The allowable stress of Douglas fir-larch Dense Select Structural is 1,900 psi.

Let's try 4 × 12s. The section modulus is a 4 × 12 is 73.83 in³ and the maximum bending stress is 120,000 ÷ 73.83 or 1,625 psi. That's acceptable.

Let's check the horizontal shear stress:

$$v = \frac{3V}{2bd}$$
$$= \frac{3 \times 2000}{2 \times 3.5 \times 11.5}$$
$$= 74.53 \text{ psi}$$

This is well below the minimum allowable stress of 85 psi.

As a final check, find the deflection of the beam:

$$D = \frac{5 \times W \times L^3}{384 \times E \times I}$$

where:
D = deflection of beam
W = total load on beam (200 × 20 = 4,000 lb)
L = span in inches (240 in)
E = modulus of elasticity (1,700,000 psi)
I = moment of inertia (415.28 in⁴)

$$D = \frac{5 \times 4000 \times 240^3}{384 \times 1,700,000 \times 415.28}$$
$$= 1.02$$

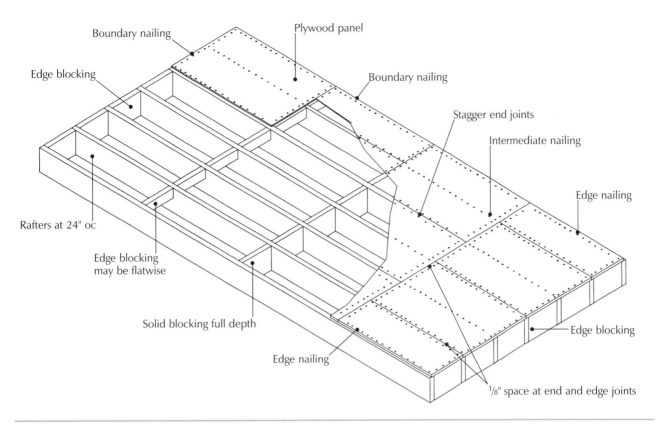

Figure 2-23 *Panel blocking and nailing*

Allowable deflection is 1.00 inch. That's close enough. You can use 4×12 Douglas fir-larch Dense Select Structural for beams spaced 4 feet on center and spanning 20 feet. The rule is, if the deflection is less than L/240, it's acceptable unless the member is supporting plaster. Then the deflection shouldn't be over L/360.

Horizontal Bracing

Brace horizontal members laterally with blocking or diagonal bracing to prevent buckling. Figure 2-23 shows typical blocking used at the ends and mid-span of floor joists. If there's no blocking and the ratio of joist span *(L)* to breadth of the joist *(b)* is greater than 20, reduce the allowable stress as follows:

Ratio (L/b)	20-30	30-40	40-50	50-60
% reduction	25%	34%	42%	50%

For example, assume a 6×14 beam spans 18 feet. The *L/b* ratio is $18 \times 12 \div 5.5$ or 39.3. You should reduce the allowable strength of the beam by about 40 percent, unless you support the top of the beam laterally to prevent buckling.

Provide lateral support for solid-sawn rectangular beams, rafters, and joists to prevent rotation or lateral displacement. If the ratio of the beam depth *(d)* to the thickness *(b)* is less than 2, no lateral support is required. If the ratio is between 3 and

4, hold the ends in place with solid blocking or bridging. If the ratio is 5, hold one edge of the beam in place for the entire length. If the ratio is 6, install bridging at third points, or no more than 8 feet on center. If the ratio is 7, hold both edges of the beam in place for the entire length.

Assume 2 × 10 joists with a 20 foot span. The *d/b* ratio is 9.5 ÷ 1.5 or 6.3. One-third of the span is 20 ÷ 3 or 6.7 feet. Therefore, use blocking or bridging at 6.7 feet on center.

Another example is a 6 × 12 beam with a 20 foot span. The *d/b* ratio is 11.5 ÷ 5.5 or 2.1, so no lateral support is required.

Designing Posts and Columns

Posts rarely fail in compression unless they are very short. Most often a post will fail in buckling. That is why the carrying capacity of a post is always based on its L/r ratio, or slenderness ratio. *L* is the vertical distance between lateral supports and *d* is the least depth or thickness of the post between lateral supports. For example, the *d* value of a 4 × 6 post is 3.5 inches and the *d* value of a 6 × 6 post is 5.5 inches.

The slenderness ratio of a solid sawn should be limited to a L/d of 50. When the L/d ratio of less than 50, the allowable compressive stress, *Fc'*, in the post is given by the formula:

$$Fc' = 0.3\,E \div (L/d)^2$$

The actual compressive stress, f_c, is:

$$f_c = P \div A$$

where:
   P = vertical load (lbs)
   A = net cross-sectional area of post (in²)
   E = modulus of elasticity (psi)

Actual compressive stress *(fc)* should always be less than allowable compressive stress *(Fc')*.

Let's look at the calculation for a typical 6 × 6 post, 16 feet long made of Douglas-fir lumber:

Modulus of elasticity *(E)*	=	1,200,000 psi
Post height	=	16'-0"
Nominal post size	=	6 × 6
Net post size	=	5¹/₂" × 5¹/₂"

Length/depth ratio (L/d) = 16 × 12 ÷ 5.5

= 35

The allowable compressive stress is:

F_c' = $0.30E ÷ (L/d)^2$

= $0.30 × 1,200,000 ÷ 35^2$

= 294 psi

Assume that the tributary roof area is 240 square feet and the dead load is:

$$
\begin{array}{rcl}
\text{Roofing} & = & 1.2 \text{ psf} \\
\text{3/4-inch sheathing} & = & 2.3 \text{ psf} \\
\text{Rafters} & = & 2.0 \text{ psf} \\
\text{Beam} & = & 0.5 \text{ psf} \\
\text{Total dead load} & = & 6.0 \text{ psf} \\
\text{Live load} & = & 16.0 \text{ psf} \\
\text{Total unit load} & = & 28.0 \text{ psf}
\end{array}
$$

$$
\begin{array}{rcl}
\text{Total load (P)} & = & 28 × 240 \\
& = & 6,720 \text{ lbs}
\end{array}
$$

Working stress is:

$$\frac{P}{A} = \frac{6720}{5.5 \times 5.5}$$

= 222.15 psi

That's less than the allowable compressive stress of 294 psi, so the post is acceptable.

Studs are essentially posts, but instead of calculating the safe loads on stud walls, use the following data to determine the approximate safe load-carrying capacities (in plf) of a 9-foot stud wall:

	Safe load (plf) at 16" oc	Safe load (plf) at 24" oc
2 × 4s	2100	1400
3 × 4s	3200	2130
2 × 6s	3200	2130

Install 2 × 4, 2 × 6 or 3 × 4 studs depending on their location, height, and loading. Use 2 × 4, 3 × 4 or 2 × 6 studs at 16 inches on center for bearing walls up to 10 feet tall. Use 3 × 4 or 2 × 6 studs on the first floor of a three-story building. Use 2 × 6 studs on unsupported 16-foot-tall walls and 2 × 8s for unsupported walls between 16 and 18 feet tall. Use 2 × 4 or 3 × 4 studs 24 inches on center for nonbearing walls up to 14 feet tall, and 2 × 6s on 24-inch centers for nonbearing walls up to 20 feet tall. Figure 2-24 shows the safe load on stud walls of various heights.

Rows of bridging Height in ft.	Size of studs at 16″ oc				
	2 x 4		2 x 6		
	1	2	1	2	4
7	2.84	3.25	4.40	5.80	5.98
8	2.20	2.67	3.41	5.56	—
9	1.74	—	—	—	—

Figure 2-24 *Safe loads on stud walls (in kips per linear foot)*

Formats for Structural Calculations

The simplest type of structural design is a conventional light-frame residential building because most of its framing members are listed in building code tables. These tables list the type, size, spacing, species, and loading condition for each kind of framing assembly. You only need to refer to a table to find the right size member for the framing system.

Building Code Tables

The 1997 UBC and the 2000 IRC include allowable span tables for:

▌ Rafters

▌ Plywood roof and floor sheathing

▌ 1-inch nominal thickness floor and roof sheathing

▌ 2-inch tongue-and-groove floor and roof decking

▌ Ceiling joists

▌ Lintels

▌ Floor joists

▌ Floor girders

In addition to the tables, many building departments provide complimentary $8^1/_2 \times 11$ detail drawings which show Type V one- and two-story wood frame buildings. These sheets also include the basic specifications required for framing. For a simple house, you can submit a floor plan, foundation plan, and elevations with an attached standard detail sheet for a permit.

Formal Written Calculations

Structural designers prepare handwritten calculations similar to the way an author writes a book. There's a title page describing the name and address of the project, with references to the building code and other authorities that govern the design and construction. The first page lists the live and dead loads plus wind and seismic factors assumed in the calculations. Of all the loads assumed, roof and floor live loads are the most important. But you don't necessarily have to use the minimum live loads given in the building codes if you anticipate heavier loads.

When you work on a design, identify the job name, page number, and date on all calculation sheets. Avoid erasing any notes. Instead, cross out errors or voided calculations. Make notes legible and easy to follow, and be sure they reflect your sequence of thought. If you select a beam that turns out to be overstressed, say so and try again with a larger member. When you select a properly-sized member, write "USE 4 × 16" GIRDER" or underline your selection.

Prepare a table of contents so that the plan checker can get to any part of the calculations quickly. Arrange your calculations starting with the roof and work down to the foundation. Using this sequence, you can follow the loads down to the footing. If you're also responsible for designing the footing, you'll be killing two birds with one stone, because you need to know the total load imposed on the footing. For example, the calculation sequence for the design of a three-story wood frame building is:

- Roof framing

- 3rd floor framing

- 2nd floor framing

- 1st floor framing

- Lateral loads (wind and seismic)

- Foundation

Make all the necessary diagrams to show loading conditions on the members. Show moment, shear, and deflection diagrams. It also helps to include a small plan of the building to locate parts you are referring to throughout the calculations. These sketches will aid the plan checker as well as the draftsperson responsible for drawing the final structural details on the plans.

Graphic Design of Trusses

Structural analysis by graphics is a simplified method for finding loads on structural members. It reduces the amount of mathematical calculations to a minimum. In fact, you can solve most structural problems with a scale, two triangles and a pencil. This method was popular long before they invented the slide rule.

Basically a line of a specific length indicates each force on a structure and direction. When all of the forces are connected end-to-end, you have a force-polygon. You use the scale of the diagram to find the length of each line of force. For example, if the scale is set at 1 inch equals 1000 pounds, you draw a 750-pound force that's ¾ inch long. The direction of the line is the same as the direction of the force. You draw a downward vertical force as a vertical line with an arrow shown at the lower

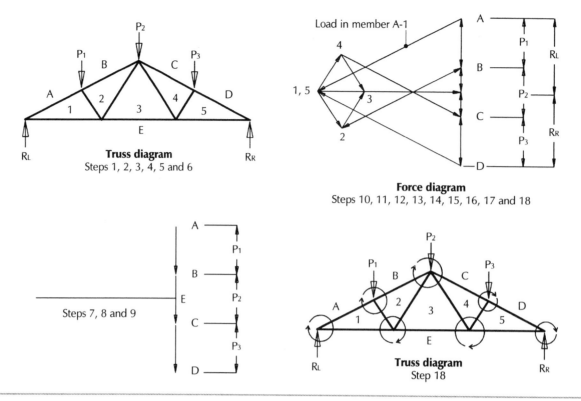

Figure 2-25 *Graphic design of a truss*

point of the line. A horizontal force to the right is a horizontal line with its arrow at the right end of the line. Diagonal forces are similarly indicated.

All forces on a structure must cancel each other out, or the structure will move or rotate. This is straight out of high school physics. That's why all force lines on a structure must close and form an equilibrium-polygon. Because of this rule, you can find the loads in the structural members of a truss when you know the outside forces. Follow these steps to determine the loads on the roof truss shown in Figure 2-25.

1) Draw the truss to scale.

2) Indicate the vertical loads at each panel point and call these loads P_1, P_2 and P_3. These are the dead and live loads on the panel points.

3) Indicate the vertical reaction at each end of the truss and call them R_L and R_R. These are the reactions at the truss supports.

4) Add letters A, B, C, D and E between the loads and the reactions. This identifies each load and reaction. P_1 is between A and B, P_2 is between B and C, and so forth.

5) Add numbers 1, 2, 3, 4 and 5 in the panels of the truss. This helps identify the truss members, such as top chord members A1 and B2 and bottom chord member E3.

6) Find the vertical load in pounds or kips at each panel point P_1, P_2 and P_3 of the truss.

7) Now draw a vertical line to scale of lengths AB, BC and CD, equal to the loads at point P_1, P_2 and P_3. I recommend you use a scale equal to 1000 pounds per inch.

8) Locate point E midway in the vertical line, assuming all vertical loads (P_1, P_2 and P_3) are equal.

9) Draw a horizontal line through point E.

10) Draw a diagonal line parallel to the slopes of the top chord from points A and D to the horizontal line.

11) Mark the point of intersection as 1 and 5.

12) Draw a diagonal line parallel to the slope of the respective web members from points B and 1.

13) Do the same for points C and 5.

14) Mark the intersections as points 2 and 4. Do the same to locate point 3 and you've completed the force diagram.

15) You can now find the load in each truss member by scaling the force diagram between points A-1, B-2, C-4, D-5, E-1, E-3, E-5, 1-2, 2-3, 3-4 and 4-5.

16) Tabulate the forces in pounds for members A-1, B-2, C-4, D-5, E-5 and E-1.

17) Find the reaction of the two supports by scaling the distance between AE and DE on the force diagram.

18) Now determine the direction of forces at each joint of all truss members. Draw a clockwise arc around each joint of the truss diagram and note the direction of the force line shown on the force diagram for that member. Force going outward from the joint means the member is in tension. Force going inward means the member is in compression. Indicate by arrows the direction of force on each member of the truss diagram.

When you order a prefabricated wood truss, give the fabricator the following information:

▌ Design dead and live loading on top and bottom chords

▌ Duration factor, or unit stress increase. The allowable stress increase in lumber and connectors due to short-time loading are:

Load	Duration	Increase
Floor load	10 years	None
Roof, snow/ice	2 months	15%
Construction	2 days	25%
Wind/seismic	1 day	33%

- Lumber specifications including the species, size, and grade of each truss member

- Force analysis, including the forces that exist in each truss member at full design load as described above

- Overall length of truss

- Description of connector plates or other types of connections

That's it for the basics of wood design. In the next chapter we'll look at the components that make up a wood-frame building.

Chapter 3

Wood-Frame Building Components

Before we study various framing systems and building components, let's look at some types of buildings as defined by the *Uniform Building Code* and the *International Building Code*. These include dwellings such as single-family houses, multifamily townhouses, condominiums, apartment buildings, and small commercial and industrial buildings.

Building Types

The *Uniform Building Code* and the 2000 *International Building Code* have revised the classifications of building occupancy. Here are the new code classifications:

Group	Occupancy
Group A, A-1, A-2, A-3, A-4	Assembly
Group B	Business
Group E	Education
Group F, F-1, F-2	Factory or industrial
Group H, H-1, H-2, H-3, H-4, H-5	Hazardous
Group I-1, I-2, I-3, I-4	Healthcare
Group M	Mercantile
Group R, R-1, R-2, R-3, R-4, R-5	Residential
Group S, S-1, S-2	Storage
Group U	Utility

Each group is subdivided and assigned a number according to relative life safety hazards. For example, R-1 includes hotels and apartment houses that have many people residing in them. R-3 includes dwellings with 10 or fewer occupants. Occupancies are classified by letter and number (R-3).

Single-Family Dwellings

Most single-family houses are classified by the UBC and the IBC as Group R-3 occupancy built with Type V one-hour construction. This means that conventional light-framed residential structures have a one-hour fire resistance rating. The one-hour rating is obtained by using exterior cement plaster and interior gypsum plaster or $^5/_8$-inch gypsum wallboard to cover and protect the wood framing in the structure.

Residential buildings classified as Type V-N construction are buildings of combustible construction. These have no fire protection rating. A private garage, attached or unattached, with wall and roof framing exposed on the inside is a Type V-N construction.

Most building departments provide sheets that describe the minimum construction requirements for one-story houses regarding fire resistance, framing, and foundations.

Townhouses

Townhouses are usually rows of single-family homes joined by common exterior side walls. Often each dwelling unit has split-level two-story living quarters over a double-car garage. Townhouse buildings are classified by the UBC and the IBC as Group R-3 occupancy and are normally built with Type V one-hour construction.

Condominium and Apartment Buildings

Condominiums and apartments are multifamily residential buildings of similar construction but with a different type of ownership. Each dwelling unit in a condominium is owned by a separate party, while all dwelling units in an apartment building are owned by a single entity. The UBC and the IBC classify both types of building as Group R-1 occupancy.

The code requires separation walls (party walls) to be relatively soundproof. Since sound travels best through solid objects, the studs between living units should be separated. You can do this by installing two parallel rows of studs or by staggering the studs. That assures that the studs of adjacent dwelling units are separated by an air space, which helps block sound transmission. Figure 3-1 shows both arrangements of studs.

Figure 3-2 shows a section through a three-story separation wall. Each wall has two rows of studs with a 1- or 2-inch air gap, and supports the joists and rafters of its respective unit.

Buildings up to three stories high may be built of Type V one-hour construction, but buildings over three stories must have an incombustible frame such as steel, concrete, masonry, or a combination of these materials. Figure 3-3 shows the frame of a three-story wood-frame building.

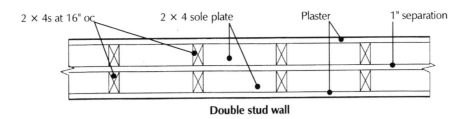

Double stud wall

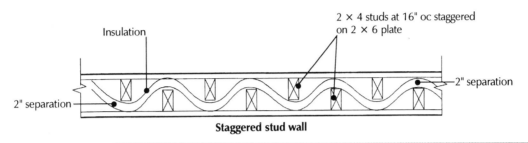

Staggered stud wall

Figure 3-1 *Typical walls between multiple dwelling units*

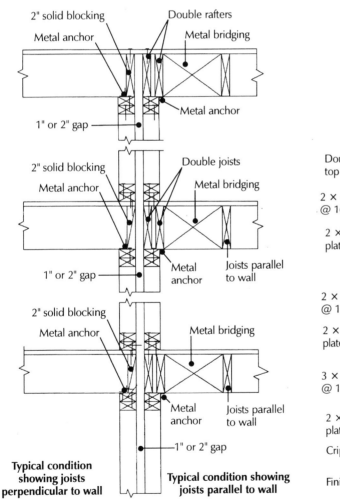

Typical condition showing joists perpendicular to wall

Typical condition showing joists parallel to wall

Figure 3-2 *Section through typical separation wall*

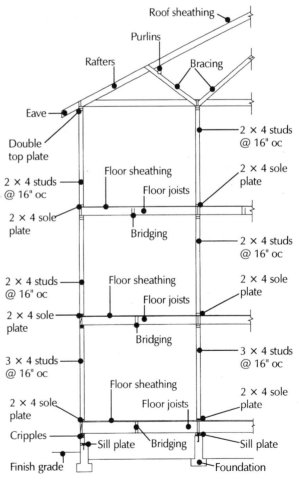

Figure 3-3 *Section through 3-story wood-frame building*

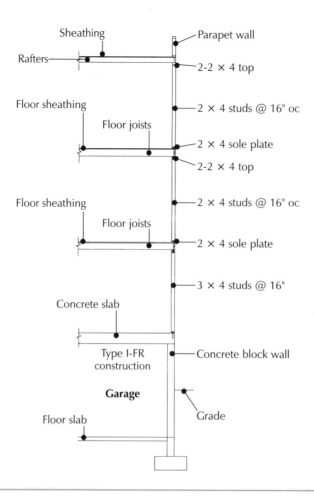

Figure 3-4 *Section through typical 3-story wood-frame building over a garage*

When a wood-frame apartment and condominium building is built over a parking structure, the parking structure must be built of incombustible Type I construction. It must also be separated from the living quarters of the building by three-hour fire-resistant construction such as a concrete slab at least 10 inches thick. Figure 3-4 shows a three-story residence built over a Type I garage.

Commercial Buildings

If the contents of a commercial building are incombustible, the UBC and the IBC classify the building as Group B occupancy. The type of framing required in a commercial building depends on the use or occupancy of the building, its size, how close it is to property lines and adjacent buildings, and the zone the building is in.

Industrial Buildings

Industrial buildings are classified as Group F1 occupancy if their contents are combustible. If the contents are noncombustible, the buildings are classified as Group F2 occupancy. Buildings in which the contents are highly flammable, explosive or hazardous to health are classified as Group H. This group is subdivided into H1 through H7, depending on the degree of hazard.

The type of construction required by the building code for an industrial building depends on the occupancy classification, building floor area and height, and how close it is to property lines and adjacent buildings.

Tilt-up Buildings with Wood Panelized Roofs

A popular type of industrial building is built with precast concrete or masonry exterior walls and a panelized (preframed) wood roof. The roof is usually made of glu-lam girders, wood purlins, subpurlins, and plywood roof panels. See Figure 3-5. Posts can be solid-sawn or glu-lam timbers. Panelized roof panels save time and labor since they're usually made at the job site, lifted, and installed on the purlins. A panelized roof forms a strong diaphragm which resists lateral loads from high winds and earthquakes.

You can use unsanded 4 × 8 to 4 × 12 foot plywood sheets supported by 2 × 4 subpurlins (stiffeners) set 16 or 24 inches on center in panelized roof construction. See Figure 3-6. The long dimension of the plywood panel can run either parallel or perpendicular to the supports, as shown in Figure 3-7.

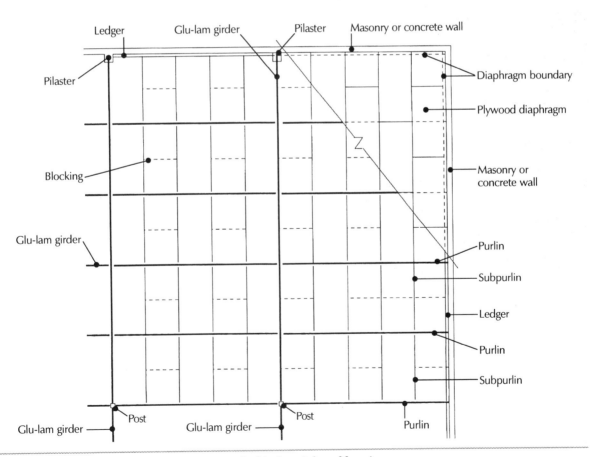

Figure 3-5 *Typical industrial roof framing*

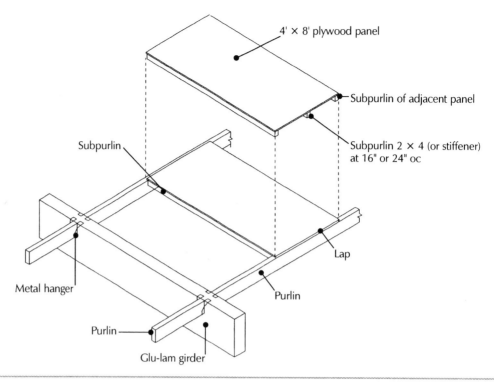

Figure 3-6 *Panelized roof*

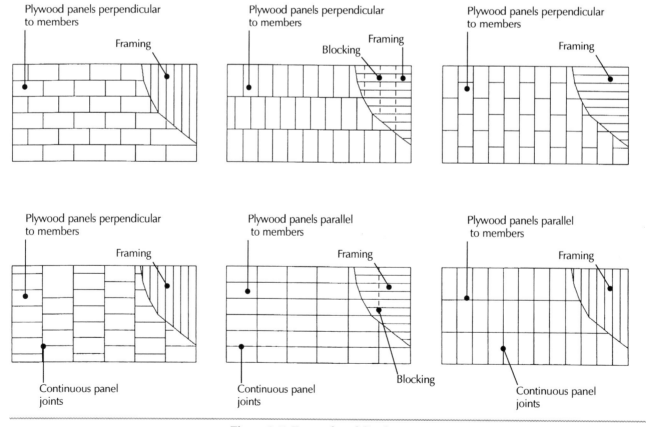

Figure 3-7 *Types of roof diaphragms*

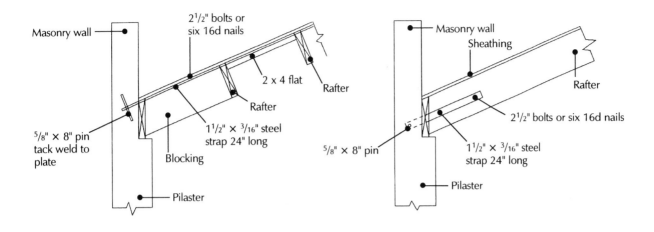

Figure 3-8 *Joist anchors in masonry walls*

Anchor the exterior walls of a structure to the roof framing system to help stabilize the walls during an earthquake. Figure 3-8 shows how to anchor sloped roof rafters to a masonry wall, for rafters that are both parallel and perpendicular to the wall. The anchor straps are attached to the side of the rafters or to the top of the blocking between rafters.

Figure 3-9 is similar, except the rafters are horizontal, not sloped. Figure 3-10 shows the plan view of the same roof framing system. Note that anchor straps are also attached to the top of the rafters or the top of the blocking between rafters. In all cases, the anchor straps are attached to the rafters with either nails, bolts or lag bolts.

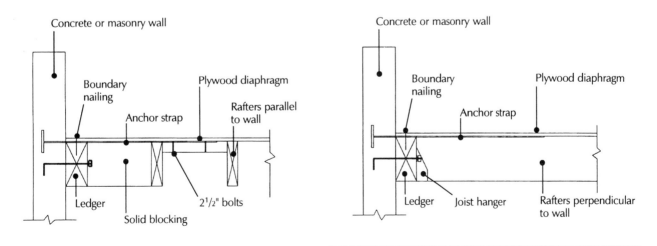

Figure 3-9 *Roof to wall seismic construction*

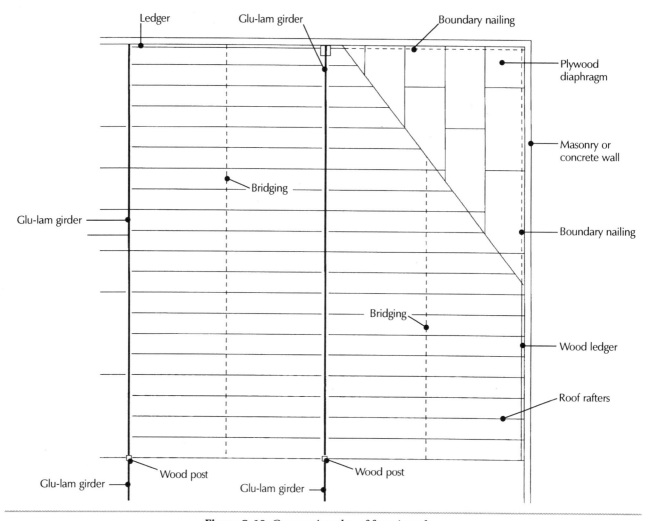

Figure 3-10 Conventional roof framing plan

Framing Systems

Now let's take a look at framing systems. These systems can be built using solid-sawn lumber or built-up or prefabricated materials such as plywood, glu-lam beams, stressed skin panels, sandwich panels, or other components. But all factory-made building components must be preapproved by the building department before you install them. The five common types of framing systems are:

- Platform
- Balloon
- Post-and-beam
- Post-frame
- Heavy timber or mill

Platform Framing

Platform framing, also called *conventional light-frame construction* or *stick construction*, is the most widely-used framing method for wood-frame buildings. A platform-framed building has floors framed in platforms with wall frames installed onto the platforms. See Figure 3-11.

Light-frame construction of unusual shape or size, and split-level structures for lateral loads in areas with a history of earthquakes or wind storms, must be designed by an engineer.

Balloon Framing

A balloon-framed building has continuous studs extending from the foundation to the roof on multistory buildings. This method was used before platform framing became popular. Balloon framing is seldom used anymore because it doesn't work with the effective shear walls and floor diaphragms which are needed to resist earthquakes and strong winds. Also, long studs are expensive and difficult to obtain.

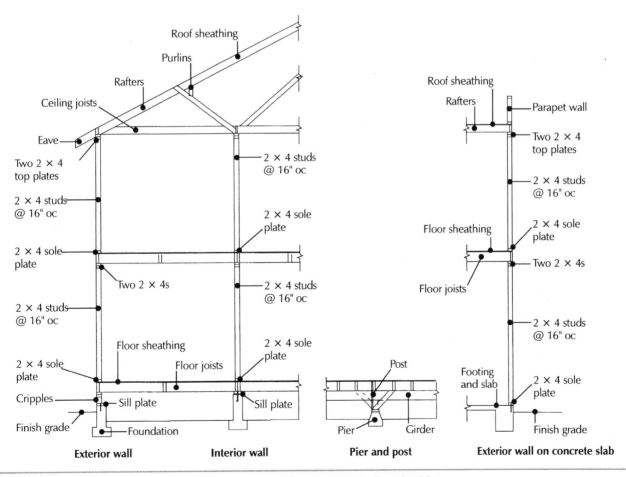

Figure 3-11 *Typical two-story platform building*

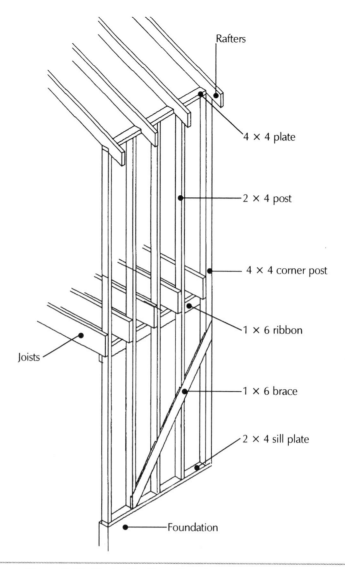

Rafters

4 × 4 plate

2 × 4 post

4 × 4 corner post

1 × 6 ribbon

1 × 6 brace

2 × 4 sill plate

Joists

Foundation

Figure 3-12 *Typical balloon framing*

Balloon framing is lighter than platform framing, but it's also less rigid. And it's more vulnerable to fire because the voids between studs are continuous from the sill plate to the roof. Note on Figure 3-12 that the intermediate floor joists are supported by 1 × 4 let-in wood ribbons nailed to the inside face of the studs.

Post-and-Beam Framing

Post-and-beam framing is framed in grids using 4 × 4s or double 2 × 4 posts. These posts are usually spaced on 4-foot centers along exterior and interior bearing walls. See Figure 3-13. The slenderness ratio of length to depth *(L/d)* of the posts must be less than 30 to give the posts adequate compressive strength. For example, the *L/d* ratio of a 4 × 4 post that's 8 feet high is 96/3.5, or 27.4. The maximum *L/d* ratio of wood posts is 50.

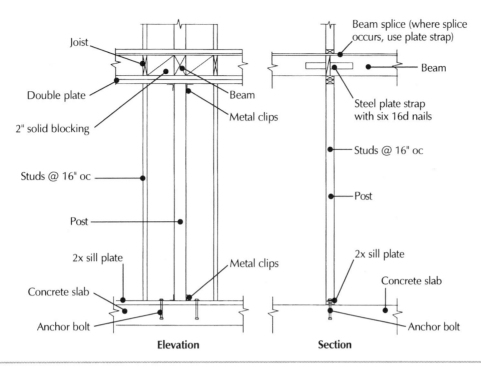

Figure 3-13 *Post-and-beam framing in stud wall*

Figures 3-14 and Figure 3-15 show the approximate load-bearing capacities of 4 × 4 to 12 × 12 posts at various heights. Note that the values shown on these tables are different because they're for different grades of lumber. The basic formula for allowable compressive stress of wood posts is:

$$F_c = 0.3\, E \div (L/d)^2$$

where:
F_c = allowable compressive stress, psi
E = modulus of elasticity of wood, psi
L = unsupported length of post, inches
d = least depth of post, inches

Capacity of typical wood posts, in pounds					
Size	**8'**	**10'**	**12'**	**14'**	**16'**
4 × 4	10,000	6,330	4,400	3,230	—
6 × 6	36,300	33,600	23,300	17,100	13,100
8 × 8	67,500	67,500	67,500	59,200	45,300

Figure 3-14 *Load-bearing capacity of typical wood posts*

Nominal size	Safe axial load (kips) Unsupported length (ft)						
	7	8	9	10	12	14	16
4 × 4	10.5	8.2	6.5	5.3	3.7	2.7	2.1
4 × 6	16.3	12.7	10.1	8.2	5.7	4.2	3.2
4 × 8	21.8	17	13.4	10.9	7.6	5.6	4.3
6 × 6	31.6	30.3	28.6	26.2	19.3	14.2	10.9
6 × 8	43	41.4	39.1	35.8	26.4	19.4	14.8
6 × 10	54.5	52.4	49.5	45.3	33.4	24.6	18.8
6 × 12	66	63.4	59.8	54.8	40.4	29.8	22.8
8 × 8	61	60.4	59.4	58.1	53.9	47.2	37.6
8 × 10	77.2	76.5	75.2	73.5	68.3	59.8	47.7
8 × 12	93.5	92.5	91.1	89	82.7	72.4	57.7
8 × 14	109.8	108.6	106.9	104.5	97	85	67.8
10 × 10	99.3	98.3	97.7	96.9	94.3	90.2	88.7
10 × 12	120.2	119	118.2	117.3	114.2	109.1	101.3
10 × 14	141.1	139.7	138.8	137.7	134	128.1	118.9
12 × 12	—	145.5	145.5	143.9	142.2	139.3	134.9

Note: Values are based on S4S dimensions.
$E = 1,600,000$ psi $Fc= = 1,100$ psi

Figure 3-15 Safe axial loads for dressed posts

The modulus of elasticity for wood can vary from 1,300,000 to 1,900,000 psi depending on the species and grade. For an exact value, do the calculation for the kind of wood you're using.

Studs or blocking installed between the posts support the interior and exterior wall covering. When the sheathing doesn't anchor the post to the top plate, provide each end with an 18-gauge metal clip angle. Anchor the bottom of each post to the sill plate with metal straps to resist horizontal and uplift forces in earthquake areas. See Figure 3-13. Install 1 × 4 let-in braces diagonally from the tops of corner posts to the sill plate to help brace the walls.

Always provide sufficient bracing to strengthen the frame against wind and earthquake forces. Minimum bracing is one diagonal brace at each corner of a wall, and intermediate bracing placed not more than 25 feet on center. For added strength, use solid masonry walls or solid exterior sheathing between large open glazed areas.

The way you anchor posts depends on the type of base you attach them to. Anchor posts to concrete with two steel angles held with anchor bolts embedded in the concrete. Anchor posts to a wood beam with nails and/or steel dowels.

Post-and-beam residential construction usually has exposed beams on the interior and exterior walls. These beams carry the ceiling and roof load. You can fill the spaces between exterior posts with fully glazed panels or a stud frame covered with exterior wall sheathing.

Post-Frame System

A post-frame building (pole building) uses a series of treated wood posts set in soil with horizontal members connecting the posts. You can build this type of building very quickly because it has no foundation. Post-frame buildings are well suited for agricultural structures because they provide large open spaces that don't require total enclosure. See Figure 3-16.

Mill Construction

Mill construction is similar to the post-and-beam system except heavy timbers are used throughout the building. Since this type of frame is more fireproof than light-frame construction, mill construction is sometimes referred to as *slow-burning construction*.

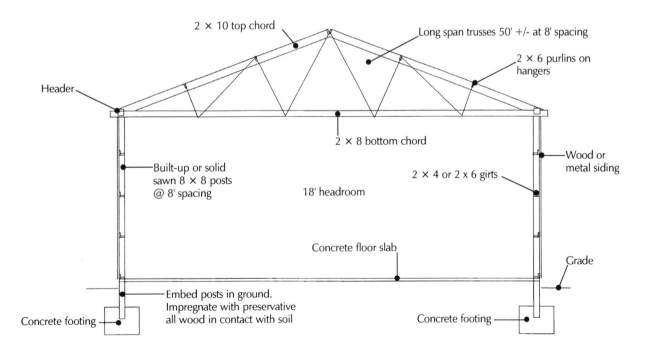

Figure 3-16 *Typical post-frame construction*

Heavy timber construction is actually more fire-resistant than unprotected steel framing because under some fire conditions, heavy timber will only scorch about $1/2$ inch deep in the same period that unprotected steel members will lose their strength. There are two reasons why it's more fire-resistant. First, timbers and planks are erected in heavy solid masses to minimize the number of ignitable projections. Second, the maximum amount of timber surface is exposed to fire sprinklers or other fire-fighting equipment.

To maximize the fire-resistance of mill construction, avoid the following:

- Concealed spaces that can't be reached by water
- Light timbers that ignite readily
- Stairway framing members under 2 inches thick
- Light-framed partitions
- Sheathing or furring

Mill buildings were popular in the early 1900s when heavy timbers were plentiful and steel shapes were expensive. These mill buildings were usually built with these features:

1) Exterior walls were made of brick masonry.

2) Roofs and floors were made of 3-inch thick planks spiked to roof and floor timbers.

3) Roof and floor bays were typically 8 to 10.5 feet wide with planks two bays long.

4) Timbers were supported on cast iron plates embedded in the masonry walls.

5) Sides of the timbers were kept $1/2$ inch clear of the brick walls to prevent dry rot.

6) All roof and floor timbers were at least 6 inches wide and 14 to 16 inches in depth.

7) Single-piece girders were preferred, although pairs of timbers were sometimes spliced with bolts.

8) Columns and posts were at least 6 inches wide. Sometimes $1^{1}/2$-inch diameter bore holes were drilled in the columns with $1/2$-inch diameter lateral vent holes near the top and bottom to allow drying.

9) Unseasoned timbers weren't painted or varnished for at least three years to prevent dry rot.

Today, the UBC and the IBC require that columns in a heavy timber building be solid-sawn or glu-lam wood at least 8 inches wide. The floor framing must be at least 6 × 10s, and roof framing must be at least 6 × 8 members. Roof and floor decks may be constructed of 3-inch thick splined or tongue-and-groove decking.

Conventional Light-Frame Buildings

There are five major building systems used in conventional light framing:

▌ Foundation

▌ Floor framing

▌ Wall framing

▌ Ceiling framing

▌ Roof framing

Foundation Framing System

The foundation framing system is made up of wood foundation cripple walls, posts, pier caps, bracing, and floor girders. Foundation cripple walls are sometimes used in pier-and-beam construction and include the sill plate, top plates, cripple studs, and diagonal bracing. See Figure 3-17.

Framing between the foundation and the first floor is the most critical for earthquake resistance. Most failures occur in this area. Framing with cripple walls is more vulnerable than framing with floor joists supported directly by the foundation walls. In active earthquake zones, make sure the framing is well anchored to the foundation to resist both horizontal and vertical seismic forces. Figure 3-18 shows two foundation walls that have been retrofitted with steel straps and plates: one cripple wall and one without cripples.

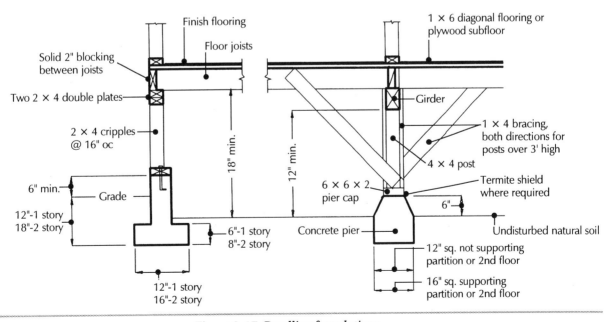

Figure 3-17 *Dwelling foundations*

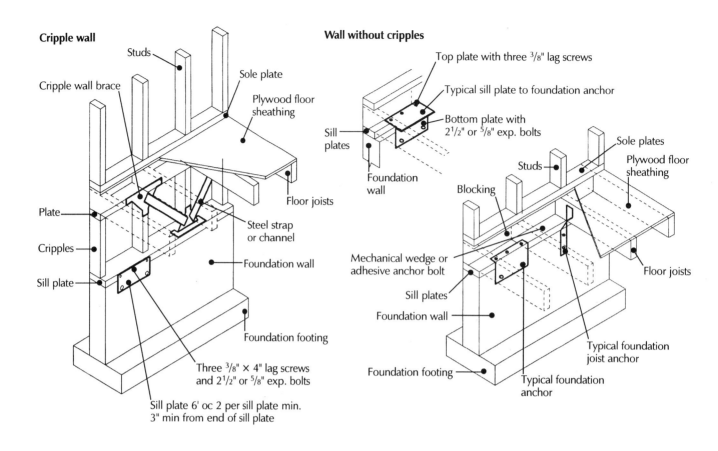

Cripple wall

Studs

Cripple wall brace

Sole plate

Plywood floor sheathing

Plate

Cripples

Sill plate

Floor joists

Steel strap or channel

Foundation wall

Foundation footing

Three ³/₈" × 4" lag screws and 2¹/₂" or ⁵/₈" exp. bolts

Sill plate 6' oc 2 per sill plate min. 3" min from end of sill plate

Wall without cripples

Top plate with three ³/₈" lag screws

Typical sill plate to foundation anchor

Bottom plate with 2¹/₂" or ⁵/₈" exp. bolts

Sill plates

Foundation wall

Blocking

Studs

Sole plates

Plywood floor sheathing

Mechanical wedge or adhesive anchor bolt

Sill plates

Foundation wall

Foundation footing

Floor joists

Typical foundation joist anchor

Typical foundation anchor

Figure 3-18 Retrofitted walls

Cripple Walls

Frame a cripple wall that's 14 to 48 inches high with studs the same size as the studs used above the cripple wall. Use solid blocking between cripple studs if the cripple wall is less than 14 inches tall. Frame cripple walls over 4 feet tall as you would frame the first floor. You can brace cripple walls with blocking or plywood sheathing.

Sill Plates

Anchor sill plates (footing sills or mud sills) to the foundation with at least ¹/₂-inch diameter anchor bolts 10 inches long embedded at least 7 inches into concrete. The anchor bolt spacing depends on the size of the anchor bolt. Space ⁵/₈-inch anchor bolts a minimum of 6 feet on center, and ¹/₂-inch anchor bolts 4 feet on center. Also, set the first and last anchor bolts within 1 foot from each end of the sill plate. See Figure 3-19. Make sure each section of sill plate has at least two anchor bolts. You can use Phillips red heads to anchor interior partitions to floor slabs.

Sill plates should be in direct contact with the top of the foundation wall, with no gaps. If necessary, use cement grout to provide a level and uniform bearing surface under sill plates. Install pressure-treated lumber, foundation-grade California redwood, or foundation lumber red cedar over interior and exterior foundation walls.

Posts

Install posts, pier caps, and diagonal braces over concrete piers that support floor girders. See Figure 3-17. Make sure each post is straight and vertically aligned, since any eccentricity may cause it to buckle. Posts are usually 4 × 4s bearing on 2 × 6 × 6 pressure-treated Douglas fir or redwood pier caps. Some builders set the caps on asphalt-saturated felt to separate the cap from the pier. Install pier caps at least 6 inches above the soil. Diagonally brace posts over 3 feet long to joists or girders in two directions with 1 × 4 braces.

Piers

Use concrete piers at least 12 inches square if they don't support partitions, and at least 16 inches square if they do. Install continuous footings under partitions that support a second floor.

Girders

Floor girders are large beams that support floor joists. They may be 4 × 4, 4 × 6, or 4 × 8 Douglas fir or southern pine. See Figure 3-20 for allowable spans of various girder sizes.

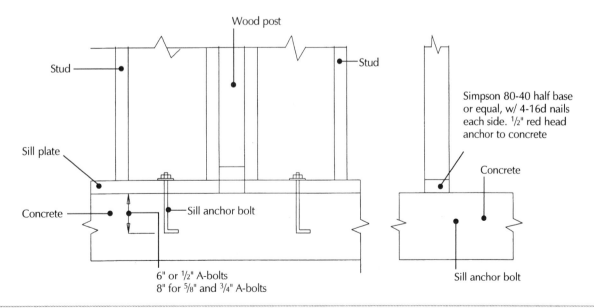

Figure 3-19 *Sill anchorage*

Girder size (in)	Spacing (ft)	Supporting partitions (ft)	Not supporting partitions (ft)
4 x 4	6	4	4
	8	3	3
4 x 6	6	7	7
	8	6	6
4 x 8	6	8	9
	8	6	7
(Based on Douglas fir larch, No. 2 or better)			

Figure 3-20 *Allowable spans for floor girders*

Girders are usually supported in pockets blocked out of the exterior concrete foundation walls and by wood posts at the interior of a building. Leave a $1/2$-inch clearance at the end of a girder and the concrete surface so the girder can rotate. This also helps keep moisture in the concrete from getting into the end of the girder. Toenail girders to each post with six 10d common nails or use galvanized steel clip angles, anchors, or T-straps. Set the top of a girder at the same height as the top of the foundation sill plate. Butt the end joints of girders over the centerline of the supporting posts. Install girders at least 1 foot above the soil.

Make sure all wood in contact with concrete or masonry within 18 inches of the ground is pressure treated, foundation grade California redwood, or foundation lumber red cedar.

Ventilation and Access

Provide at least 2 square feet of net underfloor ventilation area for each 25 linear feet of exterior wall on three sides of a building and along interior continuous footing walls. Net area is the clear area between the wires of the screen. Also install an 18- by 24-inch access opening (scuttle hole) to the underfloor space.

Floor Framing System

A floor framing system is made up of floor joists, blocking, and the subfloor. See Figure 3-17.

Joist size (in)	Spacing (in)	Supporting partitions	Not supporting partitions
2 × 6	12	10'11"	10'11"
	16	9'6"	9'11"
	24	7'9"	8'6"
2 × 8	12	14'5"	14'5"
	16	12'7"	13'1"
	24	10'3"	11'3"
2 × 10	12	18'4"	18'5"
	16	16'1"	16'9"
	24	13'1"	14'4"
2 × 12	12	22'4"	22'5"
	16	19'6"	20'4"
	24	15'11"	17'5"
(Based on Douglas fir-larch No. 2 or better and a floor live load of 40 psf. If 50 psf, reduce spans by 10%.)			

Figure 3-21 *Allowable spans for floor joists*

Floor Joists

A floor joist assembly is floor joists, blocking, or bridging. If you use untreated floor joists, be sure they're at least 18 inches above the soil. See Figure 3-21 for allowable spans of floor joists.

You can refer to Table 23-IV-J-1 in the 1997 UBC and Table R502.3.1 in the 2000 IRC for the allowable span of floor joists spaced 16 inches on center. The values are based on a 40 psf live load, 10 psf dead load, and a maximum deflection of $^1/_{360}$ of the span. The table also lists maximum spans for joists spaced 12 and 24 inches on center. Maximum spans are also given for dry lumber of varying stiffness (with a modulus of elasticity between 800,000 and 2,200,000 psi).

Although they're not always shown on the plans, you must install bridging or blocking to prevent the joists from buckling or twisting. See Figure 3-22. Bridging can be made of wood or metal. Generally you run the joists on the short span to reduce the depth and cost of the joists. This also reduces the deflection and makes the floor stiffer.

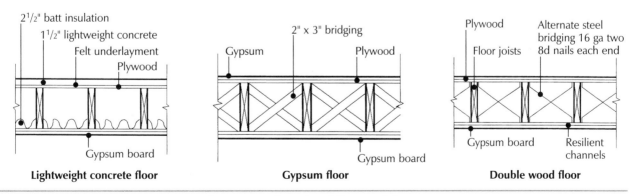

Figure 3-22 *Sound and fire-rated floors*

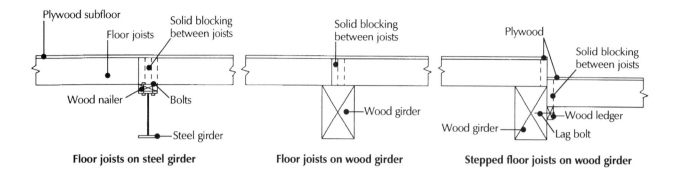

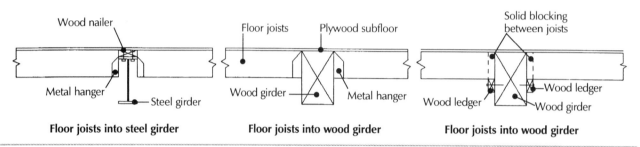

Figure 3-23 *Girder and joist connections*

In conventional light framing, joists usually rest on top of girders and lap about 4 inches over them. You can also install joists with their top surfaces at the same level as the top of a girder by suspending them with metal hangers or setting them on 2 × 4 wood ledgers nailed to the side of the girder. See Figure 3-23.

When well-seasoned timbers or glu-lam girders support unseasoned floor joists with hangers, always set the top of the girder about $1/2$ to $5/8$ inch below the top of the joists to allow for the joists to shrink. Otherwise, there will be a hump in the floor.

Use the following procedure to install floor joists:

1) Begin at the corner of the building and place joists on sill plates and floor girders. Space joists 16 or 24 inches on center since most floor panels are 4 × 8 feet and the edges must be centered over the supporting joists. Set each joist with the crown side up to help compensate for deflection. Also, install each joist with the larger knots up, since the bottom of a joist tends to split when loaded. Install double joists under any bearing partition that runs parallel to the joists.

2) Toenail the end joists and header joists (band or ribbon joists) around the perimeter of the sill plate using 8d common nails on 6-inch centers. Nail through the header joist into the end of each joist with two 16d common nails. Lap the ends of joists over girders at least 4 inches. Toenail joists to the header and girder with three 8d common nails, then face-nail the lap with three 16d nails.

3) Install rows of 2-by blocking between joists at a maximum of 8 feet on center and over the girders. For a stiffer floor system, install blocking at all unsupported edges.

Subfloor

You can use plywood, strand board, fiberboard, particleboard, planking, or square-edged or tongue-and-groove boards for subfloor material.

Plywood — Plywood is now the most common type of structural subfloor. There's less labor involved and you can install finish flooring such as carpet or resilient floor tile directly to the plywood, as long as you nail the plywood to blocking along unsupported edges. You can glue or nail plywood panels to the floor joists, or for greater stiffness, use glue and nails. The plywood thickness can range from $5/16$ to $7/8$ inch, depending on the joist spacing. APA-rated Sturd-I-Floor is a specially-designed combination subfloor-underlayment. It has a smooth surface for carpets and is resistant to concentrated and impact loads.

Plywood strength depends primarily on the net thickness and grade of the wood used in the plywood, so plywood panels are stamped with a "panel span rating" such as 32/16, 40/20, or 48/24. The first number tells you the maximum span of the panel over rafters and the second is the maximum span over floor joists. Figure 3-24 shows ratings of the plywood index system. Panel span ratings are based on the following conditions:

- The panel span must be continuous over at least two supports.

- The face grain of the panel must be installed perpendicular to the direction of the joists.

- The plywood grade must be at least C-C, C-D, Structural I, or Structural II.

- The floor live load is assumed to be no more than 40 psf, and the total load no more than 55 psf.

- The plywood panels are fastened with 8d nails spaced 12 inches on center to all supports.

Panel span rating	Thickness (in)	Span (in)	Span with unsupported edges (in)
12/0	0.3125	12	12
16/0	$5/16$, $3/8$	16	16
20/0	$5/16$, $3/8$	20	20
24/0	$3/8$, $7/16$, $1/2$	24	20 - 24
24/16	$7/16$, $1/2$	24	24
32/16	$1/2$, $15/32$, $5/8$	32	28
40/20	$19/32$, $5/8$, $3/4$	40	32
48/24	$23/32$, $3/4$, $7/8$	48	36

Figure 3-24 *Allowable spans for plywood panels*

Thickness (in)	Index	Roof, blocked (in)	Roof, unblocked (in)	Floor, unblocked (in)
$3/8$	24/0	24	20	——
$1/2$	32/16	32	28	16
$5/8$	42/20	42	32	20
$3/4$	48/24	48	36	24
$1^1/8$	Gr. 1 or 2	72	48	48

Figure 3-25 Allowable spans for plywood roof and floor sheathing

Figure 3-25 shows the allowable span for plywood roof and floor sheathing.

Start the first row of 4 × 8 plywood panels at the same corner that you started the joist layout, setting the panel edges flush with the edges of the header and end joists. Place the long dimension of the panels perpendicular to the joists. Cut panels to fit where joists lap at girders. Stagger the panel joints and nail the panels to the joists with 6d common nails for $^{15}/_{32}$-inch panels, and 8d nails for $^{19}/_{32}$- or $^{23}/_{32}$-inch-thick panels. Attach 2 × 4 scabs to the joists with 10d nails if a panel edge misses a joist. Maintain $^1/_8$-inch clearance between panel ends to allow for expansion. Set nails a minimum of $^3/_8$-inch from panel edges. Drive nails 6 inches on center along panel edges and 12 inches on center at intermediate supports.

2.4.1 Plywood Panels — You can get seven-ply plywood panels $1^1/_8$ inches thick with a front and back veneer $^1/_{10}$ inch thick made specially for post-and-beam construction. These special plywood panels are called 2.4.1 panels. They're made with tongue-and-groove or square edges. At a 4-foot span, these panels can carry a uniform load of 99 psf and a concentrated load of 385 pounds with 1/360 deflection. This type of panel provides a stiff floor system when you block the edges with 2 × 4s.

Oriented Strand Boards — Oriented strand boards (OSB or strand boards) are engineered wood products used for subfloors and roof sheathing. OSB panels are made from sawmill waste such as bark, saw trim, and sawdust. The mixture is blended with a resin binder and arranged in three layers compressed and fused under heat and pressure. The strands (or reconstituted wood fibers) in the top and bottom layers are oriented along the length of the panel face while the core strands are randomly arranged.

Strand boards are rigid and stable and have a consistent composition with no knot holes, voids, splits, or checks. OSB panels are $^1/_4$ to $1^1/_8$ inches thick. A standard size panel is 4 × 8 feet, but you can special-order panels up to 4 × 24 feet. The panels are manufactured with square or tongue-and-groove edges.

Install strand board subfloors with glue and screw them to the supports to prevent squeaking. Apply the glue in a continuous line over the supports when you install square-edged boards, or in the groove of tongue-and-groove panels. If you nail square-edge panels to joists, keep the nails at least $^1/_2$ inch from the edge of the panel. You can use 6d ring- or screw-shank nails or 8d common nails. You can also use staples if they have a $^3/_8$-inch crown and $1^5/_8$-inch leg.

Fiberboard Decking — Fiberboard is a fibrous-felted panel made from wood or cane fibers bonded with a synthetic resin. It comes $^3/_8$ and $^1/_2$ inch thick and has a smooth finish on both sides. Fiberboard is less dense than hardboard. You can use it in horizontal diaphragms or subfloors the same way as you would plywood. Drive 11-gauge roofing nails at least $^3/_8$ inch from the edge of the panel. Space the nails no more than 6 inches on center. Use edge blocking under panels spanning more than 12 inches. A $^1/_2$-inch-thick panel supported on 16-inch centers can carry a 110 psf live load. If it's supported on 24-inch centers, it can carry a live load of only 40 psf. You can also use these panels on walls and roofs.

Particleboard — Particleboard is made from a combination of wood particles and wood fibers bonded with synthetic resins. Particleboard panels come from $^1/_2$ to $^3/_4$ inch thick. You can install $^1/_2$-inch particleboard over a 16-inch joist spacing, while the $^3/_4$-inch particleboard can span 24 inches. Be sure to install the panels continuously over at least two supports and support the edges with blocking unless they're tongue-and-groove.

Tongue-and-Groove Boards — Before plywood became popular, 1 × 6 tongue-and-groove boards were the material most widely used for interior decking. It's still used for exterior decking and where labor costs are very low. One-inch-thick tongue-and-groove boards can span 24 inches as a floor deck, while 2-inch tongue-and-groove decking can span 4 feet. Don't use tongue-and-groove decking as a seismic diaphragm unless you install it diagonally because it has a low shear strength. Installed diagonally, the board sheathing serves as a subfloor as well as a horizontal diaphragm that helps resist earthquake forces and racking. Face-nail the boards to each joist with two 8d common or box nails.

Underlayment

Install underlayment over the subfloor to provide a smooth surface for floor covering materials such as carpet or tile. When you install underlayment, use a minimum $^3/_8$-inch tongue-and-groove finish flooring, $^{25}/_{32}$-inch wood strip flooring, or a minimum $^1/_4$-inch plywood underlayment installed over a tongue-and-groove or edge-blocked plywood subfloor. You can use particleboard, gypsum, lightweight concrete, and other materials for underlayment. You can kill two birds with one stone by installing $1^1/_8$ inch 2.4.1 plywood as a combination subfloor-underlayment system.

Hardwood Finish Flooring

The quality of hardwood flooring varies greatly, depending on the wood species. The common types of hardwood include red or white oak, maple, chestnut, birch, beech, hemlock, cherry, and walnut. Some finish flooring is made of softwoods, such as pine or fir.

Wood density helps determine the durability of hardwood flooring. The specific gravity of such hardwoods as white oak is about 0.68 and red oak is 0.63. Softwoods like Douglas fir and pine have a specific gravity of only 0.50.

Hardwood species are classified in a variety of ways by different agencies and associations. Each grade has a different resistance to wear and tear, moisture absorption, and deformation. Some of these classifications are:

- Premium
- Select or better
- Rustic
- Choice
- Natural or better
- Clear
- No. 1 and No. 2 common
- No. 1 common and better

Wood grading also depends on the direction of grain, and whether it comes from heartwood or sapwood. The grain direction relative to the wood strip is an important factor in its wearing ability. This is because wood grain that's perpendicular to the wearing surface has the strongest resistance to abrasion and is less likely to split or crack.

Flooring may be unfinished or prefinished with a permanent protective coating which resists moisture absorption. Some types of hardwood are impregnated with acrylic to preserve their moisture content and improve their durability.

Installing Hardwood Finish Flooring — The method you use to install a hardwood floor will affect how well the floor will wear, especially when the floor is over a crawl space that's exposed to moisture, as shown in Figure 3-26. If wood flooring gets wet, it expands, which often results in buckling, warping, cupping, and twisting. Even moderate absorption of moisture from the air can cause boards to press against one another as they swell. Excessive pressure can crush wood fibers and cause cracking. When the flooring dries, each board shrinks, leaving open cracks between the strips. A floor will usually dry most in winter when the building is heated and the average humidity is probably lower than when the building was built. To reduce this problem, some flooring manufacturers leave a small gap in the groove of tongue-and-groove boards to allow for expansion. See Figure 3-27. You can also get flooring with a wide

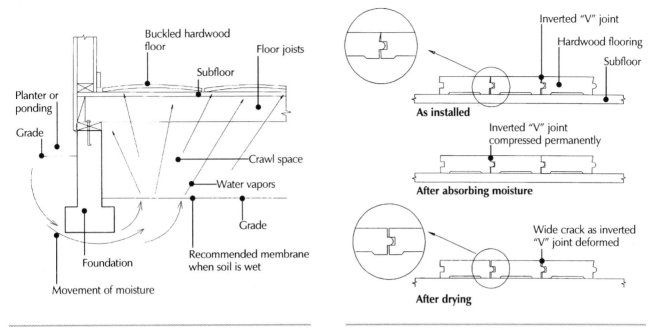

Figure 3-26 *Crawl space with moisture*

Figure 3-27 *Deformed joints in a hardwood floor*

groove along the bottom side of each board, as shown in Figure 3-28, to help prevent cupping. Some of the common defects in hardwood flooring are:

▮ Warps (parallel to the length of the board) caused by excessive moisture

▮ Open joints caused by excessive drying

▮ Cupping (parallel to the width of the board) caused by excessive moisture

▮ Loose tiles caused by inadequate nailing or adhesive

▮ Uneven surface caused by inadequate underlayment nailing

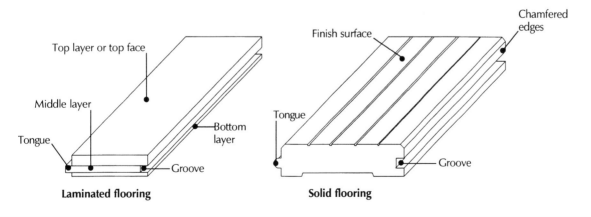

Figure 3-28 *Typical hardwood flooring sections*

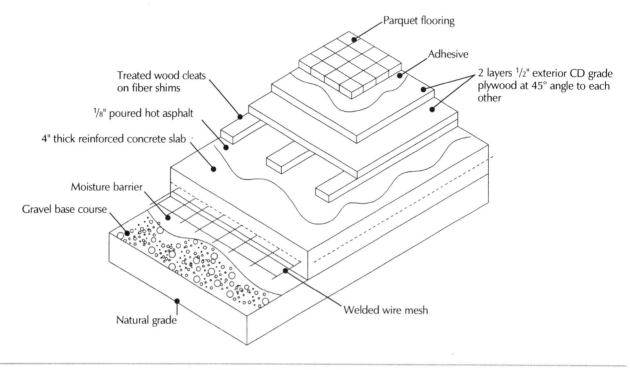

Figure 3-29 *A hardwood floor installation system*

- Cracks caused by excessive drying

- Movement caused by inadequate nailing or adhesive

- Buckling caused by excessive moisture

- Bowing caused by uneven expansion or shrinkage

- Splitting caused by sudden shrinkage

- Swelling caused by excessive moisture

- Shrinking caused by excessive drying

You can also have a problem when you install hardwood flooring directly on top of a slab on grade because moisture from the soil can come up through the slab. This is often a problem in low-cost construction. To protect the flooring, use the method shown in Figure 3-29, which has the following layers of materials:

- Cleats set over a waterproof membrane cemented directly to the concrete floor with asphaltic mastic

- Plywood nailed to wood cleats or sleepers

- Hardwood flooring blind-nailed to a plywood deck

Figure 3-30 shows another way to install a hardwood floor with spaced cleats. These cleats provide air space to help prevent the wood from absorbing moisture from the floor slab.

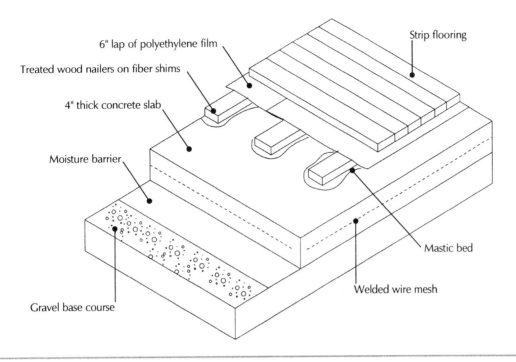

Figure 3-30 *Another hardwood floor installation system*

Figure 3-31 shows how to install hardwood flooring with and without a plywood subfloor. Make sure the subfloor and slab are dry before you install the hardwood. Use a moisture meter to measure the moisture content of the subfloor.

Here are some more recommendations to help prevent hardwood flooring problems:

▌ Finish concrete floor slabs within a maximum tolerance of $1/8$ inch per 10 feet.

▌ Thoroughly clean the surface of the slab before you apply the adhesive.

▌ Seal the slab with a primer before you apply the adhesive.

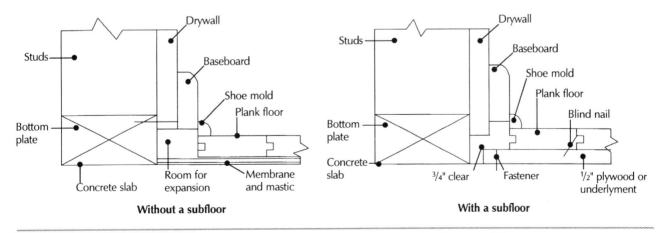

Figure 3-31 *Hardwood flooring installed with and without a subfloor*

- Install strip flooring over treated 2 × 4 wood sleepers attached to the slab with metal clips or adhesive.

- Make sure each flooring strip is nailed to at least two sleepers.

- Allow a 1/2-inch gap for expansion at all vertical surfaces such as walls.

- Make sure all the sleepers are in place before delivering the flooring, and don't deliver the flooring on a rainy day.

- Be sure the plaster and masonry walls are dry before the flooring is delivered.

- Allow recently manufactured flooring to attain equilibrium moisture content of your locale three to six weeks before it's delivered to the job site.

- Open the flooring bundles and spread the strips so that all the surfaces are exposed to air for at least four days before you install them. This allows the flooring to reach moisture equilibrium with the air before it's laid.

- Watch out for areas with excess moisture such as green concrete or wet subfloors.

- Provide some type of separation between the concrete and wood flooring using plywood, sleepers, resilient insulation board, or other similar materials.

- To be sure the slab is dry, check the moisture content using the rubber mat test or polyethylene film test. Place the rubber mat or film on the concrete surface, sealing the edges and leaving it in place for 24 hours. If water accumulates under the mat or film, moisture is getting into the concrete slab.

Wall Framing System

A wall framing system uses studs, top and bottom plates, posts, and sheathing. It also includes important secondary members for framing door and window openings, fire blocking, and bracing. Let's look at these items in detail.

Sole Plates

The sole plate (also called the bottom, lower, or toe plate) helps support the studs and connects the wall to the floor. See Figure 3-32. Install a sole plate that's as wide as the studs it supports. If you're going to nail plywood sheathing to the plate, use a 3-inch-thick plate and be sure to drive sheathing nails at least 1/2 inch from the edge of the plate.

The load-carrying capacity of stud walls is limited by the sole plate's compressive strength perpendicular to the grain rather than the strength of the stud, because sole plates can crush when overloaded.

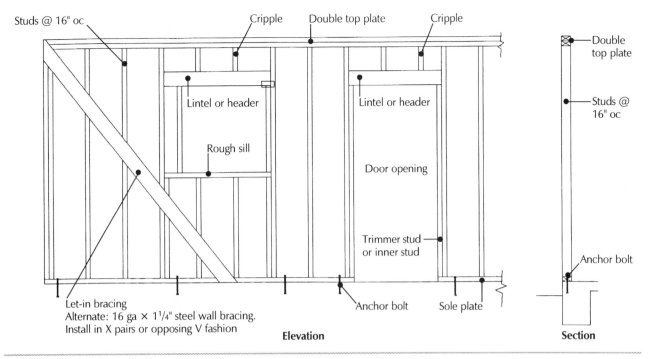

Figure 3-32 *Stud wall*

Studs

Studs support the ceiling and roof frames and provide a nailing surface for wall finish materials. Be sure the ends of studs are square with the plates so they have full bearing without gaps. Studs should run continuously without splices.

You can toenail studs to the sole plate with four 8d nails. Or you can end-nail the studs with two 16d nails, provided the sheathing extends down to the sole plate. In high-wind or seismic areas, provide additional anchorage using 1-inch-wide 18-gauge steel straps nailed at every other stud. Straps designed for this are made by Simpson Strong-Tie Company and Harlen Metal Products, Inc.

Install 2 × 4, 2 × 6, or 3 × 4 studs depending on their location, height, and loading. Use 2 × 4, 3 × 4, or 2 × 6 studs 16 inches on center for bearing walls up to 10 feet tall. Use 3 × 4 or 2 × 6 studs on the first floor of a three-story building. Use 2 × 6 studs on unsupported walls up to 16 feet tall, and 2 × 8s for unsupported walls between 16 and 18 feet tall. Use 2 × 4 or 3 × 4 studs 24 inches on center for nonbearing walls up to 14 feet tall, and 2 × 6s on 24-inch centers for nonbearing walls up to 20 feet tall.

The approximate safe load-carrying capacities (in pounds per linear foot) of a 9-foot-high stud wall are:

2 × 4s at 16 inches oc	2100 plf
3 × 4s at 16 inches oc	3200 plf
2 × 6s at 16 inches oc	3200 plf

The capacity of studs spaced on 24-inch centers is about two-thirds of studs set on 16-inch centers.

You can install double studs or staggered studs to provide soundproof separation walls between family units in condominiums and apartment buildings. Look back to Figure 3-2 for typical separation walls.

Install 3-inch-wide studs between adjacent plywood panels on shear walls. Also, keep the line of nails a minimum of $1/2$ inch from the edge of the stud to prevent splitting. You can nail a pair of 2 × 4s together to fabricate the 3-by member.

To accommodate piping or conduit, cut notches in studs no more than one-fourth their depth. Or as an alternative, drill holes up to $1^1/4$ inches in 2 × 4 studs or 2 inches in 2 × 6 studs.

Blocking

Install blocking to prevent studs from twisting or buckling and to provide a backing for nailing the edges of plywood, drywall, or particleboard panels. It'll save time if you know in advance where the panel edges will be so you can install the blocking as you frame the studs into the wall. If the edges of two adjacent panels are nailed to the blocking, install 3-inch blocking. Also, stagger the nails in adjacent panels to avoid splitting the blocking. For stud walls over 10 feet tall, install fire blocks at mid-height in the walls.

Corner Posts

Install corner posts at all exterior building corners. Use a solid 4 × 4 or a pair of 2 × 4s plus filler blocks. Figure 3-33 shows three ways to make a corner post. Nail the boards together using 16d nails 12 inches on center, plus three nails at each block.

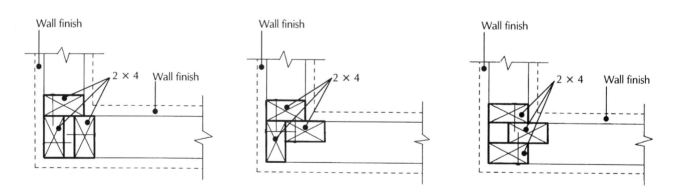

Note: Nail corner studs together to form a single unit usind 16d nails at 12" oc staggered. At least one 16d nail in each filler block through stud.

Figure 3-33 Corner post assemblies

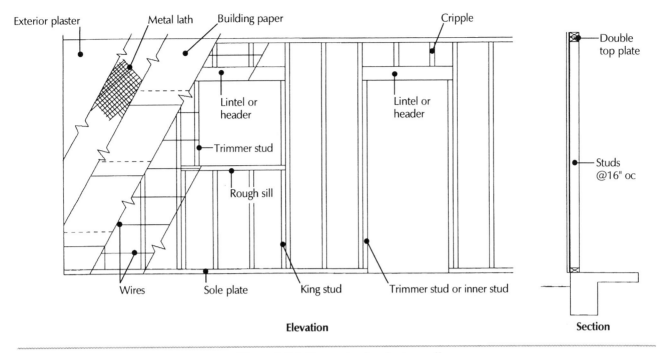

Figure 3-34 *Elements of a plaster wall*

Double Top Plates

Top plates hold the top of a wall frame together and help support ceiling joists and rafters. Top plates are usually a pair of 2-by boards the same width as the studs. Lap the uppermost member over the corners and partitions. Also, anchor the top plate to principal intersecting partitions. When you make a splice, position the butt of the lower member over a stud, and the upper member at least 32 inches beyond the butt of the lower member. End nail the lower member to the stud with two 16d nails. Nail the upper member to the lower member with 16d nails on 16-inch centers and two 16d nails at the butt of the board. When plates are cut more than one-half of their width for piping or ductwork, reinforce the plates with 18-gauge steel straps.

Beams and Headers

You can use solid continuous headers instead of double top plates. Use metal corner ties, lag screws, or other suitable fasteners to connect headers to corners and to bearing partitions. Make sure that jamb studs supporting headers at openings are designed for the load they will carry. See Figure 3-34.

Let-in Bracing

Structurally, you don't need to brace a stud wall if you sheath it with plywood or similar material, or boards installed diagonally. Nevertheless, you normally install diagonal bracing before the sheathing to keep the wall straight and plumb while you put the frame up. Braces may be 1 × 4s let into (cut into) studs, or steel straps. Install at least one brace at the end of each wall. Don't space bracing more than 25 feet

Type of sheathing	Height/width ratio	Allowable shear (psi)
1″ straight boards	2:1	50
2″ straight boards	2:1	40
Diagonal boards	3½:1	300
Lath and plaster	1:1	90
Metal lath and plaster	1:1	90
Gypsum wallboard	1:1	30
Plywood	2:1	75% of values in Table 23-I-J-1, 1994 UBC

Figure 3-35 *Height to width ratio of various types of sheathing*

apart. Install braces from the top to the bottom of a wall at a 45- to 60-degree angle with respect to the horizontal. See Figure 3-32.

Nail braces to each stud and plate with two 8d nails at each connection. If an opening is at or near a corner, install a full-length brace as close to the opening as possible. Don't use knee bracing from the top plate to the upper part of the end stud. You can install K-bracing, but a plywood sheet provides the sturdiest wall.

Properly installed let-in 1 × 4 braces can resist a maximum horizontal wind or earthquake force of 1000 pounds parallel to the wall.

Sheathing and Shear Walls

Sheathing may be made of straight or diagonal boards, plywood, gypsum board, particleboard, strand board, or other engineered wood products.

Sheathing serves as a shear wall to resist earthquake and wind loads. You can install wall sheathing to one or both sides of the studs. The height to width ratio of vertical shear walls is limited according to the material you use. See Figure 3-35. These values were based on the following conditions:

▌ All panel edges are backed with at least 2-inch-wide wood framing.

▌ Plywood panels are installed either horizontally or vertically.

▌ Nail spacing is 6 inches on center along intermediate framing members for ³/₈-inch plywood installed with face grain parallel to studs.

▌ Nail spacing along the panel edges is 2 to 6 inches on center, depending on the plywood thickness and the required shear strength.

▌ Nail spacing is 2½ inches on center at panel edges and 12 inches on center along intermediate framing for other conditions and plywood thicknesses.

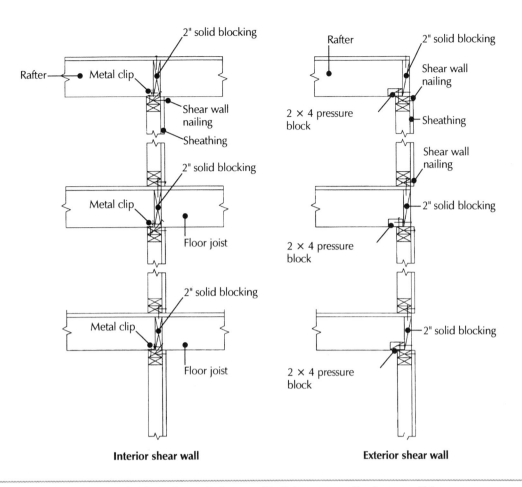

Interior shear wall   **Exterior shear wall**

Figure 3-36 *Sections through shear walls (joists perpendicular to walls)*

Shear strength of a wall is also dependent on the type and thickness of sheathing and the nailing pattern. The most critical nailing areas of a shear wall are along the wall perimeter, plywood panel edges, and intermediate framing members. The size and location of anchor bolts, hold-downs, and tie straps are also very important to the strength of the shear wall.

When you stack shear walls one above the other, nailing between shear walls is very important. The seismic forces must be carried down from the roof to the foundation. This is shown in Figures 3-36, 3-37, and 3-38.

Tests made by National Lumber Manufacturers' Association on 8 × 8 foot plywood shear wall specimens showed what happens at ultimate loading. Figure 3-39 shows how these tests were set up. Here are results of some of the tests:

Plywood on one side only:

Horizontal force, P = 7,500 lbs
Allowable shear = 200 plf
Load factor = 4.8

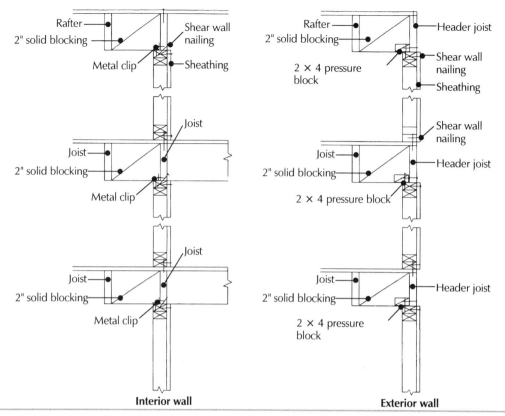

Figure 3-37 *Sections through shear walls (joists perpendicular and parallel to walls)*

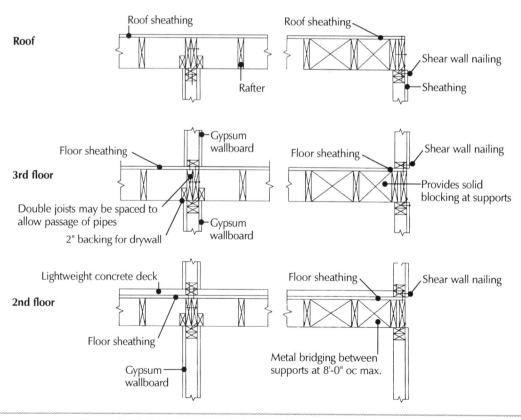

Figure 3-38 *Sections through shear walls (joists parallel to walls)*

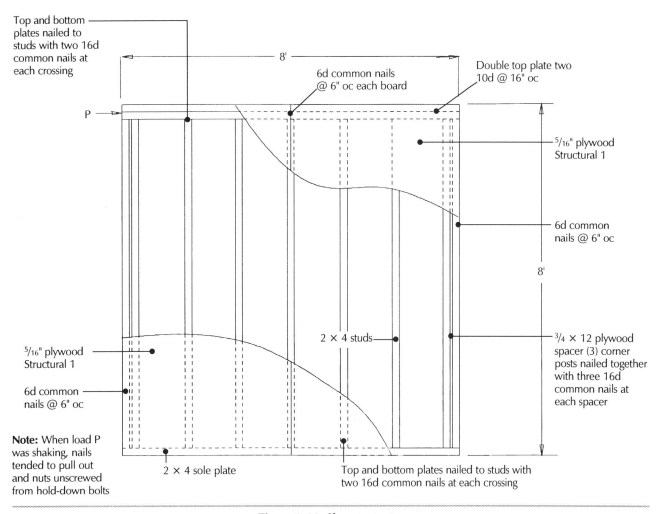

Top and bottom plates nailed to studs with two 16d common nails at each crossing

P →

5/16" plywood Structural 1

6d common nails @ 6" oc

Note: When load P was shaking, nails tended to pull out and nuts unscrewed from hold-down bolts

8'

6d common nails @ 6" oc each board

Double top plate two 10d @ 16" oc

5/16" plywood Structural 1

6d common nails @ 6" oc

8'

2 × 4 studs

3/4 × 12 plywood spacer (3) corner posts nailed together with three 16d common nails at each spacer

2 × 4 sole plate

Top and bottom plates nailed to studs with two 16d common nails at each crossing

Figure 3-39 *Shear test setup*

Plywood on both sides, boundary nails at 6 inches on center:

Horizontal force, P = 12,800 lbs
Allowable shear = 400 plf
Load factor = 4.0

Plywood on both sides, boundary nails at 2½ inches on center:

Horizontal force, P = 20,000 lbs
Allowable shear = 900 plf
Load factor = 2.0

Note: The load factor is the reciprocal of the safety factor between the ultimate strength and the allowable strength of the wall. For example, if an 8-foot-long wall failed when a horizontal force of 7500 pounds was applied to the top of the wall, this is equivalent to 7500/8 or 938 pounds per linear foot of wall. If the allowable shear strength is 200 plf, then the load factor is 938/200, or about 4.8.

When the walls failed, the plywood buckled, nails pulled out of the studs, nails pulled through plywood, and the plywood was crushed. Here's what the tests revealed:

■ Wall sheathing made of diagonal boards was four to seven times as stiff and seven to eight times as strong as horizontal boards.

■ Using three or four nails instead of two nails to attach horizontal boards to studs gave little improvement in the stiffness or strength of the wall.

■ Using three or four nails instead of two to attach diagonal boards increased the stiffness of the wall 30 to 100 percent.

■ Using 10d nails instead of 8d nails to attach horizontal boards increased the stiffness of the wall 50 percent and the strength 40 percent, but there was little improvement in walls with diagonal boards.

■ Unseasoned horizontal boards lost about 50 percent of their stiffness and 50 percent of their strength compared to seasoned lumber.

■ Diagonal 2 × 4 corner braces cut between studs made walls 60 percent stiffer and 40 percent stronger than without bracing.

■ Diagonal 1 × 4 let-in corner braces under sheathing made horizontally sheathed walls $2^1/_2$ to 4 times stiffer and $3^1/_2$ times stronger than without bracing.

■ Closely-spaced window and door openings reduced the stiffness of a horizontally sheathed wall by 30 percent and the strength by 20 percent. Diagonally sheathed walls lost 63 percent in stiffness and 50 percent in strength, but they were still much better than horizontally sheathed walls.

After studying shear wall failures in the 1994 Northridge earthquake, the Los Angeles Building Department prohibited the use of portland cement plaster or gypsum sheathing for shear walls on the ground floor of multistory buildings. Figure 3-40 shows how these materials failed during the earthquake. They also reduced the allowable shear values of gypsum sheathing by 50 percent. In addition, they recommended the maximum height to width ratios and shear strengths shown in Figure 3-35.

Door and Window Framing

Door and window framing includes cripples, trimmers, sills, and headers (lintels) as shown in Figure 3-34. Figure 3-41A gives rule-of-thumb guidance for sizing lintels, and 3-41B provides allowable spans for lintels. The strength of built-up headers made of 2-inch wide boards is about 85 percent that of a solid header built of 4-inch wide material. Use trimmer studs (jamb studs) strong enough to carry the load on the header and to resist wind load.

Set the header beam with the crown side up. Use two 2-by boards set edgewise nailed together with 16d nails on 12 inch centers. Be sure that the header has full bearing on the trimmer studs. Always frame openings over 3 feet wide with double studs.

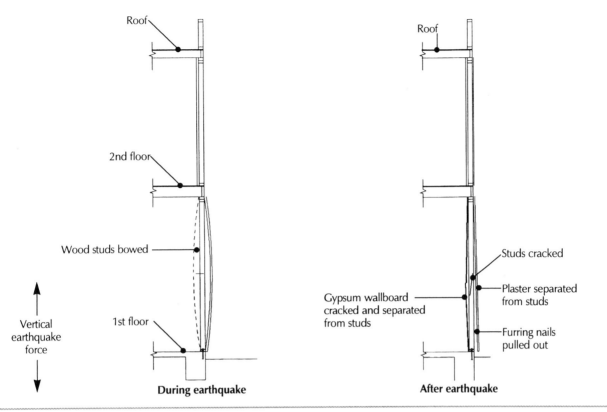

Figure 3-40 *Bowing of studs during earthquake*

A Rule of thumb for sizing lintels

Supporting	Span, L	8-0	6-0	4-0
Roof	No. of trimmers	1	1	1
	Lintel, size	4 × 8	4 × 6	4 × 4
Floor + Roof	No. of trimmers	2	2	1
	Lintel, size	4 × 10	4 × 8	4 × 6
2 floors + Roof	No. of trimmers	3	2	2
	Lintel, size	4 × 10	4 × 8	4 × 6

Note: Non-bearing wall headers should be: Opening 4-0 4 × 4 roof 4 × 6 floor

6-0 4 × 4 roof 4 × 6 floor

8-0 4 × 6 roof 4 × 8 floor

B Allowable spans for lintels

Lintel size (in)	Supporting floor, roof & ceiling	Supporting roof & ceiling only
4 × 4	3'6"	4'0"
4 × 6	5'6"	6'0"
4 × 8	7'0"	8'0"
4 × 10	9'0"	10'0"
4 × 12	10'0"	12'0"

(Based on Douglas fir-larch, No. 2 or Better. For 16-foot garage door opening in one-story attached or detached garage without ceiling, a 4 × 12 Douglas fir-larch No. 1 grade may be used.)

Figure 3-41 *Sizing lintels*

Nominal size of member	Spacing	Floor joists		Roof rafters		Ceiling joists
		With ceiling	No ceiling	With ceiling	No ceiling	
2 × 4	12			5'6"	6'8"	9'6"
	16			5'0"	5'6"	8'6"
	24			4'0"	4'6"	
2 × 6	12	10'6"	11'6"	13'0"	16'6"	18'0"
	16	9'6"	10'6"	11'6"	14'6"	16'0"
	24	7'6"	8'0"	10'0"	12'0"	
2 × 8	12	14'0"	15'0"	17'0"	21'6"	21'6"
	16	12'6"	13'6"	15'6"	19'0"	19'0"
	24	10'0"	11'0"	13'6"	15'6"	15'6"
2 × 10	12	17'6"	19'0"	21'6"		
	16	15'6"	16'6"	19'6"		
	24	13'0"	14'0"	16'6"		

(Based on plaster ceiling. Ceiling joists shall not be spaced over 16" oc unless 2 × 2 stripping at 16" oc is provided perpendicular to joists. Use Douglas fir-larch Grade No. 2 or better)

Figure 3-42 *Maximum span for floor joists, roof rafters and ceiling joists*

Ceiling/Floor Framing System

The ceiling or floor framing system in multistory buildings is made up of ceiling joists, dropped ceiling framing, soffits, floor joists, beams, headers, blocking (or bridging), and the subfloor. See Figure 3-11.

Ceiling Joists

Space ceiling joists no more than 16 inches on center unless the ceiling finish underneath will permit a greater spacing. Consult ceiling joist tables in your building code to determine the size and span of joists, or use Figure 3-42. Always check on the need for additional framing due to unusual loading conditions such as mechanical equipment mounted above the joists before you use any joist table.

Toenail ceiling joists to exterior wall plates with at least three 10d nails. Lap or butt the ends of ceiling joists over bearing partitions or beams and toenail the joists with four 10d nails.

Dropped Ceilings

Dropped ceiling framing is usually suspended from the ceiling joists or floor joists in a multistory building. You can suspend the frame using 1 x 4s or 18-gauge metal straps. When you level a dropped ceiling, never measure the ceiling height from the floor since the floor isn't always level.

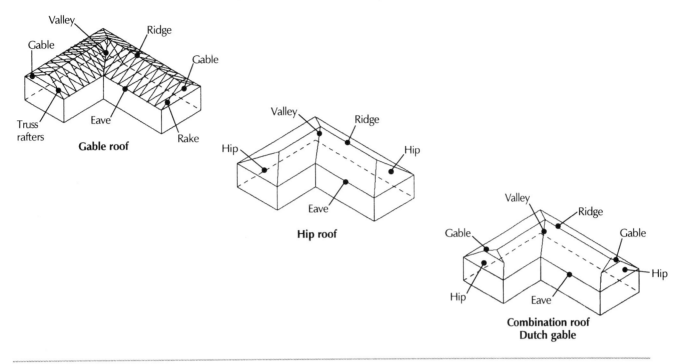

Figure 3-43 *Roof types*

Soundproof Ceilings

If a ceiling/floor system must be soundproofed, install resilient channels between the finish ceiling materials (lath or drywall) and the wood frame. Figure 3-22 shows a typical soundproof ceiling/floor system.

Roof Framing Systems

Roof framing systems may be flat, sloped or curved. Sloped roofs can be built as gables, hips, or a combination of hips and gables, as shown on Figure 3-43. A conventionally framed roof system is made up of rafters, roof bracing, ridge boards and roof sheathing.

Rafters

Rafters (or joists if they carry ceiling loads) may be solid-sawn lumber, LPI-joists, truss joists, wood I-beams, or plywood-lumber beams, as shown in Figure 1-19 back in Chapter 1. Rafters and joists may be supported by bearing walls, solid-sawn or glu-lam girders. When spans are too great for girders, you can use wood trusses.

You can place rafters 16, 24, or 48 inches on center, depending on the load and the thickness of the sheathing material, though rafters are normally spaced on 24-inch centers. See Figure 3-42. Use a continuous board from the exterior wall to the ridge board for short rafter spans. Splice boards with a gusset for longer spans. When

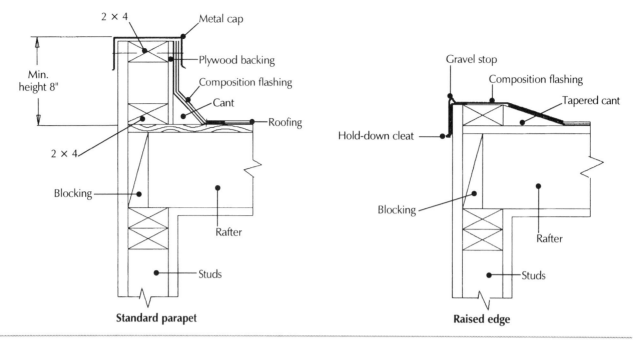

Figure 3-44 *Roof parapets*

you splice boards using a gusset, nail each rafter to the gusset with four 8d nails. Use double rafters around openings. Toenail rafters to the ridge board with at least three 8d nails.

Edges of flat roofs are usually built with a parapet or low curb to control rainwater drainage. Figure 3-44 shows construction of these two types of roof edges. I recommend the raised edge detail for curbs under 8 inches high. You need this distance to properly install cants, flashing, and metal caps. Figure 3-45 shows framing for a combination flat roof and mansard roof.

Roof Bracing

Roof bracing includes purlins, struts, uprights, and rafter ties. Install purlins horizontally under rafters where struts intersect rafters. Use purlins the same size as the rafters they support. Support purlins with braces (struts) and uprights. Notch purlins ¾ inch into braces or use 2 × 2 blocks nailed to the brace.

Bracing (struts) supporting rafters are usually 2 × 4s spaced 4 feet on center. Connect the top end of the brace to both the rafter and a purlin. Nail each brace to a rafter with three 16d nails and to a ceiling joist over a bearing wall with two 16d nails. Don't use a 2 × 4 brace longer than 8 feet. Use a 3 × 4 instead.

Install 2 × 4 or 1 × 6 rafter ties (collar beams) on 4-foot centers immediately above ceiling joists which aren't parallel to the rafters. Nail the rafter ties to rafters with four 8d nails for 1 × 6 ties or three 16d nails for 2 × 4 ties. See Figure 3-11.

Ridge Board

Use a ridge board one size larger than the rafters connected to the ridge board. A ridge board is a vertical load-bearing member when the roof slope is less than 3 to 12. Toenail rafters to the ridge board with three 8d nails.

Roof Sheathing

Roof decking (sheathing) is similar to floor decking except that it usually carries lighter loads. Construct sheathing using spaced or solid straight boards, diagonal solid boards, tongue-and-groove planking, plywood, particleboard, or other engineered wood panels.

OSB panels, for instance, are manufactured with span ratings of 40/20 and 48/24. When used for roof sheathing, the maximum span for 40/20 panels with edge supports is 40 inches, and 32 inches without edge support. The maximum span for a 48/24 panel is 48 inches with edge supports, and 36 inches without edge supports.

Fasten OSB panel edges to rafters with 8d nails 6 inches on center, and 12 inches on center along intermediate supports. Use common smooth- or deformed-shank nails for panels up to 1 inch thick. Use 8d ring- or screw-shank nails, or 10d common smooth nails for panels $1^{1}/_{8}$ inches thick. Cover roofing panels with roofing felt as soon as possible to prevent damage from moisture.

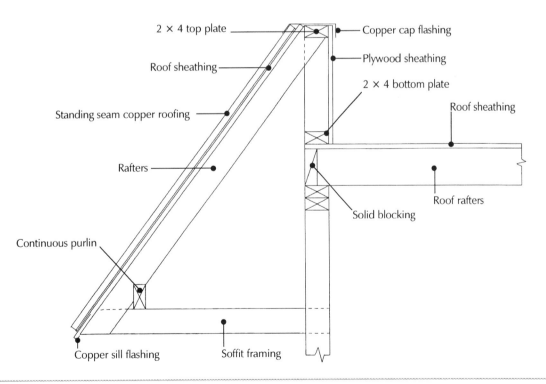

Figure 3-45 *Typical section through a mansard roof*

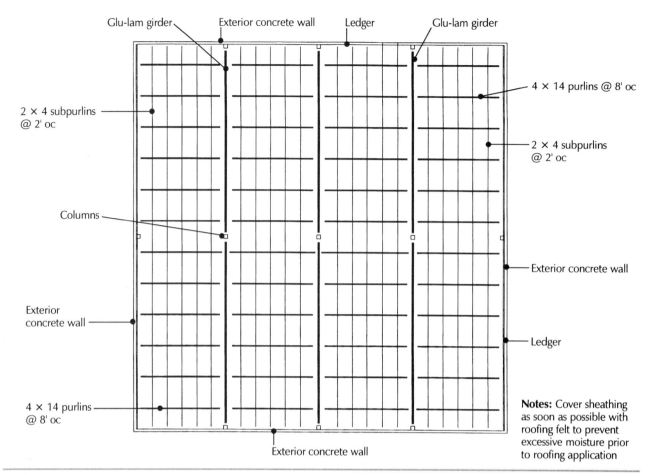

Glu-lam girder   Exterior concrete wall   Ledger   Glu-lam girder

4 × 14 purlins @ 8' oc

2 × 4 subpurlins
@ 2' oc

2 × 4 subpurlins
@ 2' oc

Columns

Exterior concrete wall

Exterior
concrete wall

Ledger

4 × 14 purlins
@ 8' oc

Notes: Cover sheathing
as soon as possible with
roofing felt to prevent
excessive moisture prior
to roofing application

Exterior concrete wall

Figure 3-46 *Roof framing plan*

Figure 3-46 shows framing for a typical flat roof on a warehouse or industrial building. This type of framing is often installed as a panelized roof system. A panelized roof is a site-fabricated roof system where plywood panels are nailed to the subpurlins on the ground, and then raised and installed on the purlins.

Chapter 4

Framing Connections

Framing systems are only as good as their connections, since most failures occur where members are joined. Every mechanical connection — nails, bolts, split rings, adhesive and hardware — must be able to safely carry loads from the top of the building to the foundation. Therefore, every connection must be at least as strong as the members you're joining.

The moisture content of the wood affects the strength of mechanical connections. For instance, nails and bolts are 10 to 20 percent weaker in wet wood than in seasoned wood. The hardness and density of the wood also affect the strength of the connections.

Nails

Nail-like fasteners typically include nails, spikes, brads, staples, and toothed steel sheet plates (gang-nails). Nail sizes or lengths are usually designated by a penny weight (abbreviated d) as shown in Figure 4-1. For example, a six-penny nail is designated as a 6d nail. Most nails in the same penny weight have the same length regardless of the size of the head or shank diameter. Some imported nails are described by their length and shank diameter instead of by penny weight.

There are some special half-size nails used to nail plywood and framing hardware. Here are some examples:

- 8d plywood nails, 1¾ inch × 10¼ gauge × $^9/_{32}$-inch head with barbed shank

- 10d plywood nails, 2 inch x 9 gauge × $^5/_{16}$-inch head with barbed shank

- 16d half length nails for hardware, 1¾ inch × 8 gauge × $^5/_{16}$-inch head

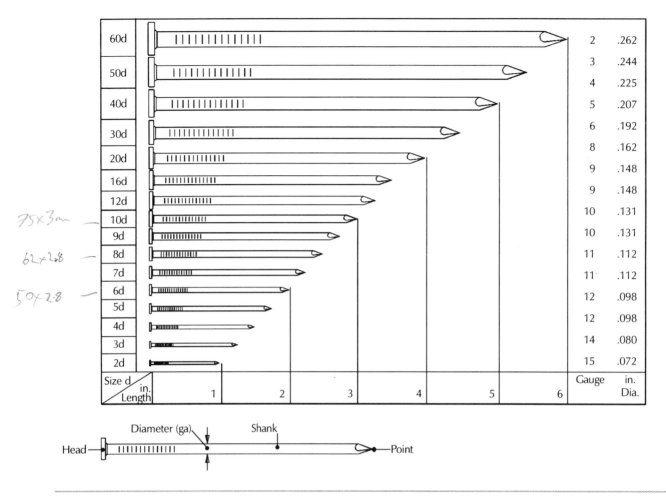

Figure 4-1 *Nail sizes and parts*

The advantage of these nails is that they don't penetrate the opposite face of 2-inch lumber. Also, they reduce labor costs, since it takes little time to drive them. Check with your local building department before using these special nails.

Common nails (steel wire nails) are the most popular. They're sold in sizes ranging from 2d (1 inch long) to 60d (6 inches long). Box nails have thinner shanks than common nails so they're less likely to split lumber. Also, you get more box nails per pound than common nails. For example, there are only 106 8d common nails per pound compared to 145 8d box nails.

Figure 4-2 shows physical data and uses for a variety of nail sizes. Figure 4-3 shows the differences between common and box nails as well as the safe lateral strength and required penetration for nails driven perpendicular to the wood grain. We'll discuss lateral strength later in this chapter. The safe load values given in Figure 4-3 are for long-term loading. Increase these values by 30 percent for wind or earthquake areas. Data is based on Table 25-G of the 1988 UBC. The table hasn't shown up in later versions.

d	Length (in)	Gauge (No.)	Head diameter (in)	Common use
2	1	15	$^{11}/_{64}$	
3	$1^1/_4$	14	$^{13}/_{64}$	lath, shingle
4	$1^1/_2$	$12^1/_2$	$^1/_4$	finish and casing
5	$1^3/_4$	$11^1/_2$	$^{17}/_{64}$	shingle
6	2	$12^1/_2$	$^{17}/_{64}$	finish and casing
7	$2^1/_4$	$10^1/_4$	$^9/_{32}$	siding
8	$2^1/_2$	$10^1/_4$	$^5/_{16}$	$^3/_4$" - 1" stock
9	$2^3/_4$	9	$^5/_{16}$	siding
10	3	9	$^{11}/_{32}$	toenailing stud
12	$3^1/_4$	9	$^{13}/_{32}$	
16	$3^1/_2$	8	$^7/_{16}$	2" stock
20	4	6	$^{13}/_{32}$	2" stock
30	$4^1/_2$	5	$^7/_{16}$	
40	5	4	$^{15}/_{32}$	3" stock
50	$5^1/_2$	3	$^1/_2$	3" stock
60	6	2	$^{17}/_{32}$	3" stock

Figure 4-2 Physical data and uses for common nails

Size	Length (in)		Gauge		Penetration (in)		Load (lbs)	
	Common	Box	Common	Box	Common	Box	Common	Box
6d	2	2	$11^1/_2$	$12^1/_2$	$1^1/_4$	$1^1/_2$	63	51
8d	$2^1/_2$	$2^1/_2$	$10^1/_4$	$11^1/_2$	$1^1/_2$	$1^1/_4$	78	63
10d	3	3	9	$10^1/_2$	$1^5/_8$	$1^1/_2$	94	76
12d	$3^1/_4$	$3^1/_4$	9	$10^1/_2$	$1^5/_8$	$1^1/_2$	94	76
16d	$3^1/_2$	$3^1/_2$	8	10	$1^3/_4$	$1^1/_2$	108	82
20d	4	4	6	9	$2^1/_8$	$1^5/_8$	139	94
30d	$4^1/_2$	$4^1/_2$	5	9	$2^1/_4$	$1^5/_8$	155	108
40d	5	5	4	8	$2^1/_2$	$1^5/_8$	176	108
50d	$5^1/_2$	—	3	—	$2^3/_4$	—	199	—
60d	6	—	2	—	$2^7/_8$	—	223	—

Figure 4-3 Properties of common and box nails in Douglas fir, larch or Southern pine

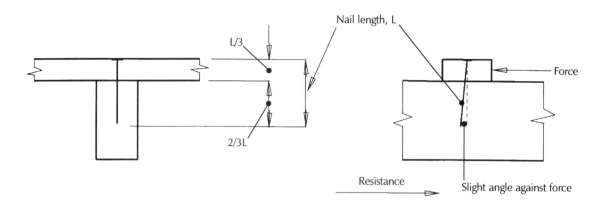

Figure 4-4 *Nail penetration*

Use a nail at least three times as long as the thickness of the wood member that it supports. Drive two-thirds of the length of a nail into the underlying wood and one-third into the piece you're fastening. See Figure 4-4. For example, use a 3-inch long 10d nail to fasten a 1-inch board. For maximum holding power, always drive nails perpendicular to the grain of the wood and at a slight angle to a line perpendicular to the wood surface.

In conventional light-frame construction, the number and size of nails for each connection are listed below:

- Joist to sill or girder: toenail with three 8d nails *60x3 mm*

- Bridging to joist: toenail each end with two 8d nails

- 1 × 6 subfloor to each joist: face nail with two 8d nails

- Wider than 1 × 6 subfloor to each joist: face nail with three 8d nails

- 2-inch subfloor to each joist: blind and face nail with two 16d nails *85x3.7mm*

- Sole plate to joist or blocking: face nail with 16d nails 16 inches on center

- Sole plate to joist or blocking on braced wall panels: use three 16d nails per 16 inches

- Top plate to stud: end nail with two 16d nails

- Stud to sole plate: toenail with four 8d nails or end nail with two 16d nails (make sure all studs have full bearing on the sole plate)

- Double studs: face nail with 16d nails 24 inches on center

- Double top plates: face nail with 16d nails 16 inches on center

- Double top plate at lapped splice: use eight 16d nails

- Blocking between joists and rafters to top plate: toenail with three 8d nails

- Rim joist to top plate: toenail with 8d nails 6 inches on center

- Top plate laps and intersections: face nail with two 16d nails: lap top plates a minimum of 12 inches

- Continuous header (two pieces): use 16d nails 16 inches on center along each edge

- Ceiling joists to plate: toenail with three 8d nails

- Continuous header to stud: toenail with four 8d nails

- Ceiling joists lap over partitions: face nail with three 16d nails

- Ceiling joists parallel to rafters: face nail with three 16d nails

- Rafters to plate: toenail the rafter assembly to the plate with three 8d nails

- 1-inch brace to each stud and plate: face nail with two 8d nails at each connection

- Up to 1 × 8 sheathing to each support: face nail with two 8d nails

- Wider than 1 × 8 sheathing to each support: face nail with three 8d nails

- Built-up corner studs: use 16d nails 24 inches on center

- Built-up girders and beams: use 20d nails 32 inches on center along the top and bottom, and two staggered 20d nails at each splice and girder ends

- 2-inch planks: use two 16d nails at each support

As a rule of thumb, use 16d nails for general framing, 8d or 10d nails for toenailing, and 6d or 8d nails for subfloor, wall sheathing, and roof sheathing. When nailing requirements aren't specified on the plans or in the specifications, follow the rules given in your local building code.

Table No. 23-II-H of the 1997 UBC lists the allowable shear in pounds for horizontal plywood diaphragms in Douglas fir-larch or southern pine framing. These values are for short-term loading due to wind or earthquake. For normal long-term loading, reduce the values by 25 percent. This table also assumes a maximum nail spacing of 10 inches along each support for floor diaphragms and 12 inches for roof diaphragms. Always check the following requirements in the job specifications:

- Plywood grade

- Common nail size

- Minimum nominal penetration

- Minimum nominal plywood thickness

- Minimum nominal width of framing member

- Allowable nail spacing at diaphragm boundaries and edges for blocked and unblocked conditions

The nailing requirements for horizontal and vertical diaphragms (shear walls) should be specified by the structural designer.

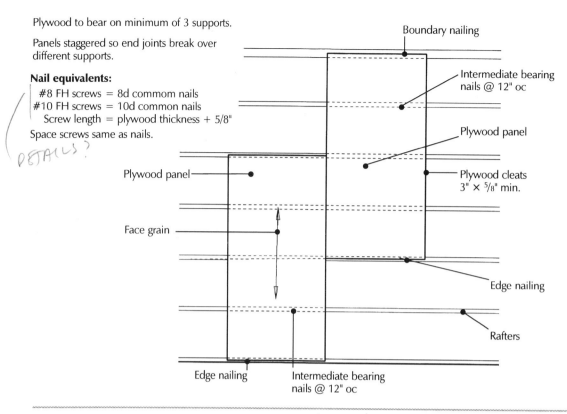

Plywood to bear on minimum of 3 supports.

Panels staggered so end joints break over different supports.

Nail equivalents:

#8 FH screws = 8d commom nails
#10 FH screws = 10d common nails
  Screw length = plywood thickness + 5/8"

Space screws same as nails.

DETAILS ?

Boundary nailing

Intermediate bearing nails @ 12" oc

Plywood panel

Plywood cleats 3" × 5/8" min.

Plywood panel

Face grain

Edge nailing

Rafters

Edge nailing

Intermediate bearing nails @ 12" oc

Figure 4-5 *Plywood panel nailing locations*

To avoid installation errors at the job site, it's a good idea to use colored pencils to divide the roof plan into separate areas according to the required nailing patterns along panel edges and intermediate panel supports. Figure 4-5 shows plywood panel nailing locations for a typical roof diaphragm. For example, a panelized roof framing system of a tilt-up warehouse as shown in Chapter 3, Figure 3-5 with 4 × 10 ledgers, 4 × 14 purlins at 8 feet on center, 2 × 4 subpurlins at 2 feet on center, and ½-inch 4 × 8 plywood panels laid in a staggered pattern, may require the following nailing:

Boundaries (at ledgers)	10d at 2 inches oc
Panel edges	
Area I	10d at 3 inches oc
Area II	10d at 6 oc
Intermediate supports	10d at 12 inches oc

Parts of a Nail

The parts of a nail include its shank, head, tip, surface finish, and composition. Shanks may be smooth or deformed with a ringed, screwed, threaded, clinched, or barbed shank. Due to friction, deformed shanks provide greater withdrawal strength than common wire nails. They can also carry greater lateral loads. The disadvantage is that some deformed shanks, especially grooved or spiral shanks, tend to break more easily, especially when the joints undergo vibration or stress reversal. Stress reversal occurs when a load or force repeatedly changes direction, such as during an earthquake or in a structure supporting dynamic machinery.

As a rule of thumb, you can use a 5d threaded nail in place of a 6d common nail, and a 7d threaded nail instead of an 8d common nail. Another rule is that you should provide two nails in 3-, 4-, 6-, and 8-inch wide boards and three nails in 10- and 12-inch wide boards.

Nails are manufactured with a variety of heads, including round, countersunk, double-headed, common flat, large flat, reinforced large flat, checkered, sinker with twin-head, pointed cone, long narrow brad and cupped brad.

The shank finish may be bright, galvanized, cement-coated, plastic (vinyl)-coated, or blued. When nails are tumbled to remove dirt and chips, they take on a bright surface. Galvanized nails are coated with zinc applied in a hot tumbler, by electrolysis, or by a mechanical process. Cement-coated nails are made by applying an adhesive in a tumbler. Cement-coated nails have almost twice the holding strength of smooth nails. An acid-etched nail has still greater power since the acid roughens its surface. Blued nails are produced by heating the nails until an oxidation layer is formed.

Some nails are made of aluminum, stainless steel, copper, brass, or Monel metal. Most nails are made from steel wire, so their shank diameters are rated in terms of wire gauge. See Figure 4-1 to compare nail gauge to shank diameter.

Double-Headed Nails

Double-headed nails (scaffold nails, staging nails, or concrete form nails) are used for temporary connections that will eventually be dismantled, such as concrete formwork. They're made in sizes from 6d to 20d. The inner head stops the nail short of full penetration into the wood, leaving the outer head exposed so they're easy to remove.

Special Nails for Plywood and Hangers

There are some nails made specially for plywood sheathing. These are short nails with thick deformed shanks. Here are some examples:

- 8d plywood nails, 1¾ inches × 9 ga × $^9/_{32}$-inch head, diamond point and barbed shank

- 10d plywood nails, 2 inches × 9 ga × $^5/_{16}$-inch head, diamond point and barbed shank

- 16d half-nail, 1¾ inches × 8 ga × $^5/_{16}$-inch head, used for purlin hangers and other metal joint devices

Hangers increase the strength of a connection, so you can use smaller nails, as shown in Figure 4-6.

Specified nail	Replacement nail	Allowable load adjustment factor
16d common	10d × 1½"	0.64
16d common	10d common	0.83
16d common	10d sinker	0.83
16d common	10d × 1½"	0.77
16d common	16d sinker	1.00

Figure 4-6 Special short nails for hangers

Special Roofing Nails

There are many types of roofing nails. Some of them are listed in Figure 4-7 by name, finish, shank length, and diameter.

Nails for Power Nailing Tools

You can use pneumatic-power tools to drive common nails ranging from 2d to 6d, spikes ranging from 3 to 12 inches long, and finish nails, brads, and staples.

Staples for power tools are stored in coils which can hold up to 180 staples. Nails are stored in magazines (sticks). Each stick can hold up to 100 6d nails.

You can adjust pneumatic hammers for the amount of nail penetration you want, depending on the lateral strength and withdrawal strength you need. Tables for lateral and withdrawal strength are usually provided by the air gun manufacturer.

Name	Finish	Shank length and diameter (in)
annular ring shank (22 to 336 rings per inch)	bright steel	¾ × 0.130 to 1 × 0.140
screw shank	aluminum	1 × 0.150
smooth shank	bright steel	⅞ × 0.105
file shank	bright steel	1 × 0.130
steep spiral screw shank	bright steel	1 × 0.140
barbed shank	galvanized	¾ × 0.145

Figure 4-7 Name, finish, shank length and diameter of common roofing nails

Some of the most common manufacturers of mechanically-powered, pneumatic, and powder-actuated power tools are:

- Air-Nail Company

- Air-Staple Company

- Grip Rite West

- Halstead Enterprises

- Hilti, Inc.

- Stanley Bostitch

- Kay Staple International

- Weyerhaeuser Company

Here are some of the common uses for pneumatic tools:

- Use framing stick nailers or utility coil nailers for framing, sheathing, roof decking, subflooring, strapping, siding, exterior decks, bridging, and to drive plain, ring, and screwed nails from 2 to 4 inches long.

- Use pneumatic staplers to do specialized work such as lathing, roofing, insulation, and finish carpentry (staples range from ½ to 2 inches long with a ½- to 1-inch crown).

- Use special pneumatic tools to drive concrete nails and spikes up to 5 inches long.

Pneumatic tools must be connected by hose to special air compressors with from 1 to 5 horsepower. Compressors are driven by electric motors or LPG engines equipped with 4- to 8-gallon capacity compressed air storage tanks.

The maximum operating speed of a pneumatic tool depends on the type and number of tools operating simultaneously. Some large compressors can service as many as four framing nailers at one time.

The Air-Nail Company has an oil-less strip nailer that can drive nails from 0.113 to 0.162 inch in diameter and 2 to 3½ inches long. See Figure 4-8.

You can also nail and staple with powder-actuated tools which use a small cartridge of gunpowder to drive the fastener into the wood or concrete.

Spikes

Spikes are long, heavy-duty nails. Ten-penny (10d) nails 3 inches long and larger are called spikes. Double-threaded spikes do the work of three ordinary spikes because their spiral threads rotate into the wood and help resist withdrawal. Predrill holes for spikes ¼ inch or larger in diameter.

Courtesy: Air-Nail Company

Figure 4-8 *Atro oil-less 16d nailer*

Nailing

Here are some things to consider when you're driving nails:

▮ Toenailed connections aren't as strong as perpendicular connections. The allowable lateral strength per nail in a toenailed joint is about 50 percent of that given in Figure 4-9.

▮ Drive toenails at no more than a 45-degree angle with respect to the grain.

▮ Drive toenails starting at a point from the end of the board approximately one-third the length of the nail.

▮ Keep nails spaced no closer than one-third the nail length, and keep nails away from the edge of the board at least one-fourth the nail length.

▮ Don't use toenails to resist withdrawal forces.

▮ The lateral strength of box nails is only about 75 percent of common nails since box nails have a smaller shank.

Wood group	Allowable lateral strength of nails by size* (lbs)					
	6d	**8d**	**10d**	**12d**	**16d**	**20d**
I	78	97	116	116	132	171
II	63	78	94	94	107	139
III	51	64	77	77	88	139
IV	41	51	62	62	70	91
*Normal installation of common, ring and screw shank nails						

Figure 4-9 *Allowable lateral strength of nails*

- Reduce the allowable lateral strength of a connection by 75 percent if you use unseasoned wood.

- Be sure to use nails long enough so at least one-half the nail penetrates the supporting member.

- Nails that split a board will do little to resist any force.

- Drive at least two nails at any connection.

You can predrill nail holes up to three-quarters the nail diameter without affecting the lateral or withdrawal strength of the connection.

Lateral and Withdrawal Forces on Nails

Nails must resist two types of forces: side grain lateral force and side grain withdrawal force, as shown in Figure 4-10. The lateral strength of a nail is determined by its resistance to a force perpendicular to its shaft. Here's the formula:

$$P_L = KD^{3/2}$$

where:

P_L = allowable lateral strength of a nail (lbs) when it's driven into the side grain of seasoned wood

K = coefficient based on specific gravity (SG) of wood:

SG	K
.29-.42	1440
.43-.47	1800
.48-.52	2200

D = diameter of shank (inches)

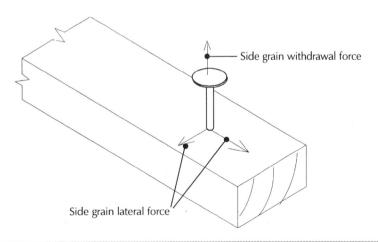

Figure 4-10 *Forces on nails*

Nails by pennyweight (d)	Lateral strength (lbs) in Douglas fir, coast grade, larch, western and Southern pine
2	27
4	42
6	52
8	65
10	78
12	78
16	90
20	116
30	130
40	147
50	166
60	185

Based on formula: P = K x D$^{3/2}$ where K = 1375   D = diameter of nail (in)
Value of K varies with different species and densities.
Note: Nail penetration should be $^2/_3$ length of nail.

Figure 4-11 *Lateral strength of nails by pennyweight*

The penetration of a nail affects the lateral load as shown by this formula:

$$P_{LI} = \frac{P_L \times X}{^2/_3 L}$$

where:

P_{LI} = limited lateral load (lbs)
P_L = allowable lateral load in optimum penetration (lbs)
X = length of shank in holding piece (in)
L = nail length (in)

Figure 4-9 lists the allowable lateral strength of nails for various wood groups. For information on wood groups see Figure 1-3 in Chapter 1. Figure 4-11 shows the lateral strength of nails by pennyweight.

The withdrawal strength of a nail is:

$P_W = K_W \times D$

where:

P_W = allowable withdrawal strength in pounds per linear inch of shank from side grain of holding piece
K_W = 295 for all grades of Douglas fir, 132 for redwood (K is based on the specific gravity of the wood, which varies in each species)
D = diameter of shank (in)

For adequate holding power, make sure a nail penetrates at least 14 times its shank diameter. For example, the minimum penetration for a 16d nail is 14 x 0.162, or about 2¼ inches. If the penetration is reduced, the holding power is also reduced. For example, a nail penetration of 8.5 times its shank diameter gives only 61 percent holding power (8.5 ÷ 14). Nails driven parallel to the grain have little resistance to withdrawal forces, so don't load a member with tension parallel to the nails holding it.

Approximate withdrawal strength values for some common nails driven into Douglas fir are:

Size	Lbs per inch of penetration
6d common	29 (not less than 12 shank diameters)
8d common	34
10d common	38

Here's a list of the safe resistance to withdrawal of common wire nails driven into a seasoned redwood member. These values are per linear inch that the nail goes into the member.

Size	Lbs per inch of penetration
6d	15
8d	17
10d	20
12d	20
16d	21
20d	25
30d	27
40d	30
50d	32
60d	35

Bolts

Machine bolts are the most common type of bolts used in heavy timber construction. They're ¼- to 1½ inches in diameter and have hexagonal or square heads. Hexagonal heads are better for power tools, while square heads are more convenient for hand tools. It's a good idea to install washers under the head and nut when a bolt bears on wood to increase the strength of a connection. Washers may be cut steel, malleable iron, or cast iron. Flat square steel plate washers with barbed corners are also made for anchor bolts in mud sills. This prevents the bolts from shifting in the hole.

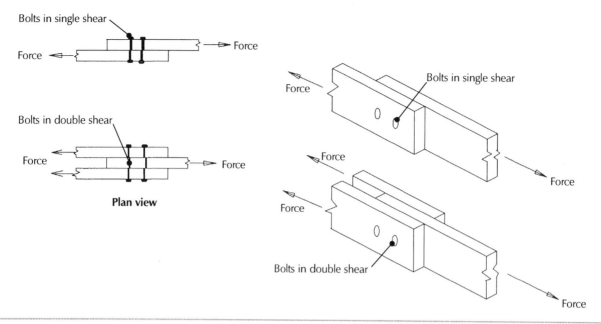

Figure 4-12 Single and double shear

The strength of a bolted connection depends on many factors, including:

▌ Bolt diameter

▌ Number of bolts per connection

▌ Single or double shear (see Figure 4-12)

▌ Thickness of wood member

▌ Use of steel side plates or washers

▌ Direction of grain in wood

▌ Species and density of the wood

You can increase the strength of most bolted connections by 25 percent if you use steel side plates. This is in addition to the increase allowed for short-term loads like wind or earthquake loads. Inspect all bolted connections of wood trusses and structural frames periodically to make sure each connection is tight, since wood shrinks as it dries.

Bolt Holes

You must predrill all bolted connections. You'll need a cutter head, pilot and bit for split-ring connectors. Space bolt holes parallel to the grain and at least 4 bolt diameters apart. Also space bolt holes at least 7 bolt diameters from the end of a member to prevent splitting. See Figure 4-13.

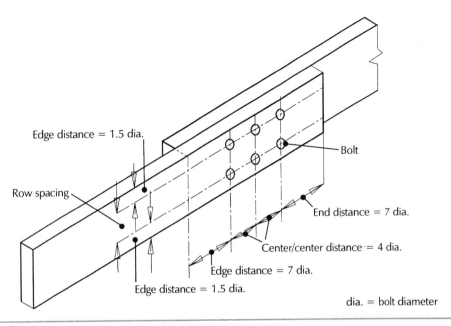

Edge distance = 1.5 dia.

Row spacing

Bolt

End distance = 7 dia.

Center/center distance = 4 dia.

Edge distance = 7 dia.

Edge distance = 1.5 dia.

dia. = bolt diameter

Figure 4-13 Bolt spacing

Lag Bolts (Lag Screws)

When you can't get to both sides of a joint, you can use lag bolts (lag screws). A lag bolt is a wood screw with a hexagonal or square head. Its diameter is $3/16$ to $1\frac{1}{4}$ inches, and it's from 1 to 12 inches long. Lag bolts don't work well under vibrating or earthquake loads.

The formula for withdrawal strength of a lag screw is:

$$P = 1800 \times \sqrt{G^3} \times \sqrt[4]{D^3} \times L$$

where:

    P = maximum withdrawal (lbs)
    G = specific gravity of the wood based on 12 percent moisture content
    D = shank diameter (in)
    L = length of penetration of the threaded portion (in)

For example, let's find the withdrawal strength of a lag bolt where the specific gravity of wood is 0.60, the bolt is 1 inch long, and its shank diameter is 0.25 inches. Then:

$$P = 1800 \times \sqrt{60^3} \times \sqrt[4]{0.25^3} \times 1$$
$$= 296 \text{ pounds}$$

Length (in)	Withdrawal resistance (lbs) by gauge number										
	#6	#7	#8	#9	#10	#11	#12	#14	#16	#18	#20
1	41	45	49	53	57	61	65	—	—	—	—
1½	62	68	74	80	85	91	97	110	—	—	—
2	—	90	98	106	114	122	130	146	160	—	—
2½	—	—	—	133	142	153	162	182	200	220	—
3	—	—	—	—	—	—	195	219	240	264	288

Figure 4-14 *Withdrawal resistance of lag screws*

where:

$G = 0.60$ (G is the ratio of density of wood to the density of water, expressed as a decimal)

$D = 0.25$ inches

$L = 1$ inch

Since most designers use tables such as Figure 4-14 to figure out lag screw strength, this formula is here just to show you how you could figure the strength of lag screws if you had to.

You must make adjustments for the strength of wood lag screws, depending on the condition of the wood, as well as the service environment. You can use these factors:

Lumber condition	% of withdrawal load	% of lateral load
Seasoned, always dry	100	100
Exposed to weather	75	75
Always wet	67	67
Metal side plates	100	100

Let's redo the previous example for a screw exposed to the weather. The withdrawal strength would be:

$$P = 0.75 \times 296$$
$$= 222 \text{ pounds}$$

You can use Figure 4-15 to find the allowable lateral load for lag bolts screwed into a member that's 1½ inch thick or thicker. Note that the allowable load depends on the wood group and the direction of the bolt with respect to the wood grain. For information on wood groups see Figure 1-3 in Chapter 1.

Figure 4-16 lists the withdrawal strength of screws according to gauge and length. Figure 4-17 shows the lateral strength of screws according to gauge.

Lag bolt diameter (in)	Bolt parallel to grain (lbs)		Bolt perpendicular to grain (lbs)	
	Wood group		Wood group	
	II	III	II	III
0.25	170	130	170	120
0.375	240	170	180	130
0.5	290	210	190	140
0.625	360	260	10	150

Figure 4-15 Lateral strength of lag bolts

Screw length (in)	Withdrawal strength by gauge number* (lbs)										
	6	7	8	9	10	11	12	14	16	18	20
1	41	45	49	53	57	61	65	—	—	—	—
1½	62	68	74	80	85	91	97	110	—	—	—
2	—	90	98	106	114	122	130	146	160	—	—
2½	—	—	—	133	142	153	162	182	200	220	—
3	—	—	—	—	—	—	195	219	240	264	288

*Screw penetration into second member is at least $^2/_3$ of screw length.

Figure 4-16 Withdrawal strength of screws

Gauge no.	Lateral resistance* (lbs)
6	51
7	62
8	73
9	84
10	98
11	111
12	126
14	158
16	194
18	234
20	276
24	374

*Screw penetration in second member is 7 times screw diameter.

Figure 4-17 Lateral resistance of screws

Drift Pins

Drift pins are similar to bolts except they aren't threaded, so the withdrawal strength depends only on friction. Drift pins are driven into predrilled holes and resist only shear or lateral forces.

Anchor Bolts

Anchor bolts are threaded bolts embedded in a concrete foundation, slab or wall. The threaded end remains exposed so you can fasten framing members to it. You can install anchor bolts before, during or after you pour concrete. If you wait until the concrete sets up, you'll have to drill a hole into the hard concrete and install an expansion bolt or similar device. You can also use an epoxy adhesive to anchor a threaded rod in the hole.

Here's how you install a mechanical wedge anchor bolt:

■ Drill the hole to the desired depth.

■ Clean the hole with a nylon brush, compressed air, or blow bulb.

■ Insert the anchor bolt, making sure it's free of dirt, oil, or grease.

■ Tighten the bolt according to the manufacturer's instructions.

■ As you tighten the bolt, the tapered shaft will pull up, wedging the clip into the sides of the hole.

To install a threaded rod using an encapsulated adhesive:

■ Drill and clean the hole as described above.

■ Insert the capsule of adhesive in the hole.

■ Attach the threaded anchor rod to a rotary hammer adapter.

■ Rotate and hammer the rod to break the capsule and mix the contents.

■ Allow the adhesive to cure around the threaded rod. Curing time varies from 12 to 24 hours, depending on the type of epoxy you use.

If you mix the adhesive:

■ Drill and clean the hole.

■ Mix the two-component adhesive and pour it into the hole.

■ Push the threaded rod into the hole.

■ Rotate the rod slightly to thoroughly coat it with adhesive.

■ Allow the adhesive to cure around the anchor rod.

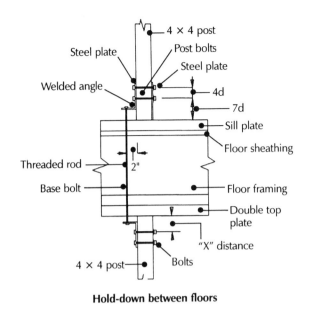

Hold-down between floors

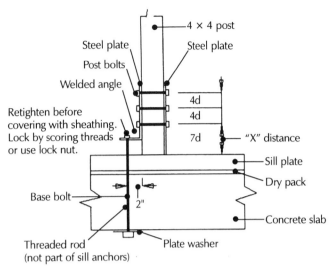

Hold-down at slab

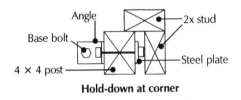

Hold-down at corner

Note: Anchor strength in concrete is dependent on corner distance and anchor center-to-center distance.

Figure 4-18 Hold-down details

Bolts used for hold-downs provide tension to the ends of a shear wall. See Figure 4-18. Drill bolt holes $1/16$ inch larger than the diameter of the bolt you use in the hold-down. Tighten the hold-down connectors just before you cover the wall frame.

The required embedment and withdrawal strength of various sized anchor bolts for hold-downs are:

Bolt diameter (in)	Embedment (in)	Tension (lbs)
$5/8$	4	1500
$3/4$	5	2250
$7/8$	7	3200
1	8	3200

Red heads manufactured by Phillips Drill Company are made with self-drilling snap-off anchors or stud anchors. A red head is a self-drilling device that becomes an anchor. Install these anchors with a motor hammer or by hand. After the hole is drilled, you insert a hardened steel plug into the anchor, then reinsert the plug and anchor into the hole. Finally, impact the device with an electric- or air-driven hammer, which expands the anchor, locking it in place.

Bolt size (in)	Tensile load (lbs)	Shear load (lbs)	Min. edge or centers (in)
$1/4$	4000	1363	$1^1/_4$
$5/16$	4425	2073	$1^1/_2$
$3/8$	6175	3443	$1^1/_2$
$1/2$	9260	6856	$2^1/_8$
$5/8$	12760	12153	$2^1/_2$
$3/4$	17675	16516	$2^1/_2$
$7/8$	19440	18816	$3^1/_2$

Figure 4-19 *Tensile and shear strengths of snap-off anchors*

The tensile and shear strengths for snap-off anchors are given in Figure 4-19. These are ultimate strength values based on 3,000 psi concrete. Use 25 percent of these values for design.

Anchor bolts are critical in earthquake areas. The 1994 Northridge earthquake led to new methods and rules regarding connections used to hold a building down to its foundation. Here are some of the new methods:

▊ Anchor the sill plate (mud sill) with ½-inch anchor bolts spaced a maximum of 6 feet on center, with a bolt 1 foot from each end of the sill plate.

▊ Always install a steel washer with each anchor bolt.

▊ Use framing anchors to connect the cripple studs to the sill plate.

▊ Use framing anchors to connect floor joists to band (header) joists.

▊ Use hold-down anchors to connect plywood shear walls to double studs. See Figure 4-18. Install hold-downs at each corner of a building. Use solid 4 × 4 or 4 × 6 posts at each hold-down unless the plans specify other sizes.

One weak point of an anchor bolt is that the nut holding down the mud sill has a tendency to loosen during an earthquake, so be sure you tighten the nut well and lock it in position.

During an earthquake, an anchor bolt can also move horizontally in an oversize hole drilled in the mud sill. This causes the bolt to transfer all of the load to one edge of the hole, splitting the wood. Also, the building may shift ¼ inch or more before the bolt begins to restrain movement. To avoid these problems, don't drill holes more than $1/_{16}$ inch larger than the bolt size. Or you can install a square-toothed steel washer embedded into the mud sill. Install these washers on top, under, or on both the top and bottom of the mud sill.

Retrofitting with Anchor Bolts

Retrofitting with new anchor bolts or threaded rods is much more difficult than placing them when the foundation is poured. There are many devices available which provide anchorage into hard concrete. Using expansion bolts in predrilled holes is the most common. Regardless of the method you select, use a metal detector to locate reinforcing bars in the concrete before you drill a hole.

Check with your local building department before using any of these anchors to see what values they allow. Use one of the following methods to install new bolts in an existing concrete foundation, wall, or slab.

Expansion Bolts in Predrilled Holes

Here's how to install a two-unit ring wedge anchor:

1) Drill a hole to the specified diameter and depth.

2) Clean the hole of dust and chips.

3) Insert head of bolt with first unit into the hole.

4) Expand the unit in position by swedging with a piece of pipe or special caulking tool.

5) Insert the second unit and expand.

6) Attach the connected item and tighten with a nut.

Two-Component Adhesive and Threaded Rod in Predrilled Holes

1) Drill a hole to the specified diameter and depth.

2) Clean and remove dust from the hole.

3) Fill the hole with epoxy.

4) Insert the threaded rod and turn slowly for full contact.

5) Allow epoxy to cure.

6) Attach the connected item and tighten with a nut.

Screw Anchors

1) Drill a hole to the specified diameter and depth.

2) Clean the hole of dust and chips.

3) Insert the screw anchor with a power tool.

4) Attach the connected item and tighten with a nut.

Lead Anchors

1) Drill a hole to the specified diameter and depth.

2) Clean the hole of dust and chips.

3) Insert the lead anchor or shield.

4) Insert the screw into the lead anchor with a power tool.

5) Attach the connected item and tighten with a nut.

Red Head Snap-Off Anchor

1) Insert the tapered end of the snap-off anchor into the chuck head attached to an impact hammer.

2) Use the impact hammer to drill into the concrete. Rotate the chuck handle while drilling.

3) Withdraw the drill.

4) Clean out the hole.

5) Insert the hardened steel cone-shaped red expander plug in the cutting end of the drill.

6) Re-insert the plugged drill in the hold and use a hammer to expand the anchor.

7) Snap off the chucking end of the anchor with quick lateral strain on the hammer.

8) Attach the connected item and tighten with a nut.

Powder-Actuated Anchors

Powder-actuated devices are popular because of the speed of installation, but they are dangerous to operate. You can use these tools for fastening 2-inch thick wood sills to concrete. The hardened steel studs are available with either a head or threaded shank.

Here are some of the regulations required by California Construction Safety Orders for the use of powder-actuated tools:

▮ You should only use powder-actuated tools that meet the design requirements in ANSI A10.3-1977 or have a state approval number.

▮ The tools can only be operated by qualified persons who carry valid operators cards for the tools.

▮ Each tool should have an operator's manual, power load and fastener chart, tool inspection and service record and service tools and accessories.

▮ You should never leave a loaded tool unattended.

▮ Don't drive fasteners into very hard or brittle materials including, but not limited to, cast iron, glazed tile, hardened steel, glass block, natural rock, hollow tile and most brick.

▮ Don't drive fasteners into concrete unless the material thickness is at least three times the fastener shank penetration.

- Don't drive fasteners into any spalled areas.

- All operators and assistants must wear eye and face protection.

- Never point loaded or unloaded tools at anyone.

- You must conspicuously post this 8 × 10-inch sign with bold letters within 50 feet of the area where the tools are being used:

CAUTION - POWDER-ACTUATED TOOL IN USE

Miscellaneous Connectors

Other miscellaneous connectors you can use for connections are split rings, toothed rings, shear plates, segmental washers, steel dowels, and self-drilling fasteners.

Split Rings

Use split rings for wood-to-wood connections. Since split rings are stronger than bolts, you can usually use a single split ring instead of a group of bolts in a connection. Place the ring into precut grooves in the contact face of the each member, fitting half the ring into each face at the point of contact. Then insert a ¾-inch diameter bolt and nut into the rings and tighten the bolt with a hydraulic press. You can get split rings 2½, 4, and 6 inches in diameter, 1 inch deep. See Figure 4-20.

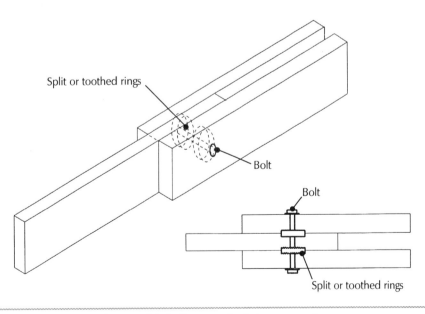

Figure 4-20 *Split or toothed ring connectors*

Toothed Rings

You can use toothed rings for wood-to-wood bolted connections. Toothed rings are similar to split rings except they have corrugated metal edges that pierce the wood faces. You place the ring between two pieces of wood and draw the wood together under pressure. Toothed rings require no power equipment so you can use them where there's no power available. Toothed rings are made 2, $2^5/8$, $3^3/8$, and 4 inches in diameter.

Shear Plates

You can use shear plates for steel-to-wood or dismountable wood-to-wood connections. A dismountable connection is one that can be adjusted by tightening to correct for wood shrinkage. You place the plates in precut grooves so the connector is flush with the surface of the wood. Then you tighten the connection with power equipment. Shear plates are made $2^5/8$ and 4 inches in diameter.

Segmental Washers

You can use segmental washers to tighten connections in shrinking heavy timber structures. Since you insert segmental washers under a nut so friction holds the segments together, you don't have to remove the nuts to add more washers. They're very handy when timber has shrunk to the point where the bolt thread isn't long enough to let you tighten it.

Steel Dowels

Steel dowels are good for anchoring a beam to the top of a post. Always use glue to help hold a dowel in a hole. Epoxy glue works best.

Self-Drilling Fasteners

You can use self-drilling fasteners to connect wood to steel. These fasteners are made of a hardened steel with a drill point. The drill point is followed by an unthreaded and then a threaded portion. The tool drills a pilot hole first, then it cuts threads into the steel plate and screws itself in. See Figure 4-21. They're made with hexagonal, countersunk, rounded, or Phillips heads. Hexagonal heads are made with an integral steel washer. You can add a neoprene washer under the steel washer to waterproof the hole.

You can use self-drilling fasteners for these connections:

- 2 × 4 wood to steel member: use a 6-20 × $1^5/16$ flat head, TEKS/4 with pilot head

- $3/8$- to $1/2$-inch thick wood siding to 12-gauge steel girt: use a 6-20 × $1^5/16$ trumpet head TEKS/3 with pilot head

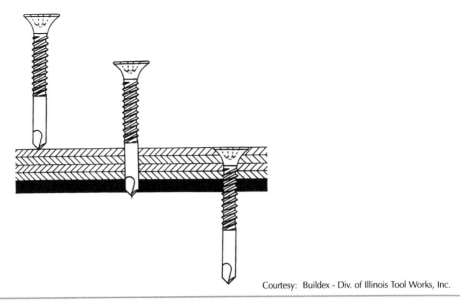

Figure 4-21 *Self-drilling fasteners*

- $^5/_8$- or ¾-inch plywood flooring to 12-gauge steel joists: use a 6-20 × $1^5/_{16}$ trumpet head, TEKS/3 with pilot head

- ¾-inch or thinner plywood to steel framing up to 0.175 inch thick: use a 10-24 × $1^7/_{16}$ inch wafer head, plymetal TEKS/3 with wings

Framing Hardware

Due to the increased demand for earthquake- and wind-resistant buildings, there are many types of metal connecting devices being used in timber framing. Building codes now require that plans show detailed information regarding the framing hardware. The lateral force system for resisting wind or earthquakes should be clearly shown on the foundation and framing plans. There should also be sufficient elevations and detailed references for all shear walls, frames, and foundation anchors.

Some framing hardware items are shown in Figure 4-22. These hardware items come predrilled or punched so you can attach them to the framing members with bolts or nails. Here's a brief description of the most common framing hardware.

Post Caps and Anchors

Post caps are made of 16-gauge steel for 4 × 4, 4 × 6, or 6 × 6 wood posts. Attach post caps to a post with 16d nails or ½-inch bolts. Column caps are made of 7-gauge steel.

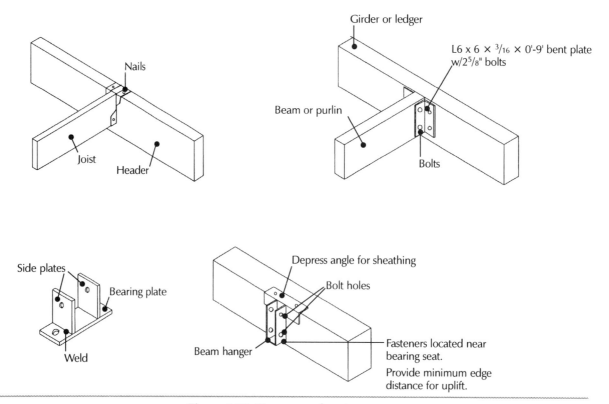

Figure 4-22 *Hangers and seats*

Post anchors are also made of 16-gauge steel for 4 × 4, 4 × 6, or 6 × 6 wood posts. Attach post anchors to a post with 16d nails or ½-inch bolts. Figure 4-23 shows a post anchor made of a ¹/₈- × 1¼-inch steel strap and a ¹/₈-inch steel plate. Be sure to set these anchors before you pour the concrete.

Dowel Plates

Dowel plates are made of ½-, ⁵/₈- or ¾-inch thick steel cut to fit the size of a wood post. The plate contains a welded ½- or ¾-inch diameter pin 8 or 12 inches long. Embed the lower half of the pin in the concrete and insert the upper half into the center of the wood post. The pin keeps the bottom of the post from shifting and the plate protects the wood from infestation.

Joist Anchors (Framing Anchors)

Use joist anchors to hold wood joists or rafters to a masonry or concrete wall. An anchor is made up of a strap and a pin which you insert through a hole in the end of the strap. Straps are ¹/₈ or ¼ inch thick, 2 inches wide and 24 to 48 inches long. The strap is punched with holes for two ½-inch bolts or six 16d nails for attachment to the joist. A ⁵/₈-inch-diameter pin 8 inches long is embedded into the center of the masonry or concrete wall. Joist anchors are usually placed 48 inches on center. These connections are very important in earthquake zones to prevent masonry or concrete walls from falling away from the building. See Figure 3-9 back in Chapter 3.

Framing anchors are used for lighter-duty connections such as studs, joists, and rafters. They're made of pressed 18-gauge galvanized steel sheets shaped for 2 × 4 subpurlins and joists. Use five 11-gauge by 1¼-inch nails in each leg of these anchors.

Joist and Beam Hangers

There are many types of joist and beam hangers. They usually hang from wood headers, ledgers, girders, or steel beams and carry the end of joists, beams, or purlins. Some are made from ⅛-inch-thick welded steel angles and bar stock. Others are made from pressed 16- to 11-gauge sheet metal. Attach them to the wood members with 16d nails or weld them to the top flange of steel beams. These hangers are made for 2 × 6 to 2 × 16, 3 × 6 to 3 × 16, and 4 × 6 to 4 × 16 wood members. See Figure 4-22.

Hold-Downs (Tie-Down Anchors)

You install hold-downs to prevent a shear wall from rotating off its foundation and to transfer shear and rotational loads between floors. One type of hold-down is shown in Figure 4-18. Be sure you accurately place the hold-downs in the foundation or slab. And be sure the foundation is heavy enough to hold the shear wall down. The anchor strength in concrete depends on the spacing of the anchors, as well as the distance from a building corner to an anchor.

Figure 4-24 shows typical schedules for hold-downs and tie straps. (Also refer to Figure 4-18.)

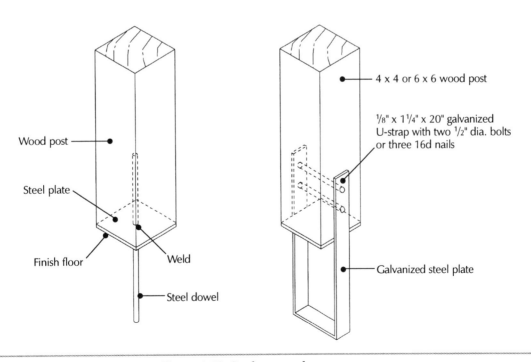

Figure 4-23 Steel post anchors

Typical schedule for post-to-post hold-down:

Mark	Type	X-distance (in.)*	Base bolt	Post bolt
1	HD-2	$4^3/_8$	1 - $^5/_8$" dia AB	2 - $^5/_8$" dia MB
2	HD-5	$5^1/_4$	1 - $^3/_4$" dia AB	2 - $^3/_4$" dia MB
3	HD-6	$5^1/_4$	1 - 1" dia AB	3 - $^3/_4$" dia MB
4	HD-7	$6^1/_8$	1 - $1^1/_8$" dia AB	3 - $^7/_8$" dia MB

*X-distance is the minimum end distance from the first bolt in the post.

Typical schedule for tie straps in post-to-post connections:*

Mark	Type	Material	Width (in)	Nails
1	TS-1	20 ga Gl	$2^5/_{16}$	10 - 16d
2	TS-2	16 ga Gl	$2^5/_{16}$	14 - 16d
3	TS-3	16 ga Gl	$2^5/_{16}$	20 - 16d
4	TS-4	12 ga Gl	$2^1/_{16}$	15 - 16d
5	TS-5	12 ga Gl	$2^1/_{16}$	21 - 16d
6	TS-6	12 ga Gl	$2^1/_{16}$	25 - 16d

*All end posts for strap connection shall be 4 x 4.

Typical schedule for hold-downs to a slab:

Mark	Type*	X-distance (in)	Base bolt	Post bolt
1	HD-2	$4^3/_8$	1 - $^5/_8$" dia AB	2 - $^5/_8$" dia MB
2	HD-5	$5^1/_4$	1 - $^3/_4$" dia AB	2 - $^3/_4$" dia MB
3	HD-6	$5^1/_4$	1 - 1" dia AB	3 - $^3/_4$" dia MB
4	HD-7	$6^1/_8$	1 - $1^1/_8$" dia AB	3 - $^7/_8$" dia MB

*For HD-2 and HD-5, use 4 x 4 posts minimum; for HD-6 and HD-7, use 4 x 6 posts.

Notes:

Mark	=	symbol on the plans at hold-down (HD) and tie strap (TS) locations
Type	=	Strong-Tie hold-down and tie straps
X-distance	=	minimum distance between the first post bolt and the nearest end of the post, as shown in Figure 4-18
Base bolt	=	number and diameter of anchor bolts, as shown in Figure 4-18
Post bolt	=	number and diameter of machine bolts in the post, as shown in Figure 4-18

Figure 4-24 Typical schedules for hold-downs and tie straps

Mud-Sill Anchor Singleside (MAS)

Mud-sill anchors are manufactured by Simpson as a substitute for the conventional ½-inch anchor bolt used to hold down mud sills. The MAS anchor is attached to only one side of the mud sill, so it's faster and easier to install than the conventional anchor bolt. It eliminates drilling the mud sill and does away with mislocated anchors bolts. You can install these units before the concrete is poured by nailing them to the forms. Or you can set them after you pour the concrete by inserting them in the wet semiplastic concrete. It's easier to finish the edge of slabs because there are no anchor bolts sticking out. All you have to do is slip the mud sill under the anchor, and then nail it with two 10d nails to the top of the mud sill and four 10d nails to the side of the mud sill. The uplift strength of a MAS unit is 990 pounds. See Figure 4-25.

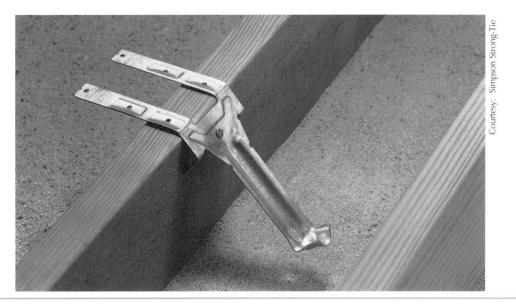

Courtesy: Simpson Strong-Tie

Figure 4-25 *Mud-sill anchor (MAS)*

Flat Straps

Flat straps are used as tie-downs in shear walls. One type is made of 16-gauge steel 1½ inches wide and 24 inches long. The straps are punched for 16d nails at each end. Always make sure the listed strength of these straps published by the manufacturer is approved by your local building department.

Steel Wall Bracing Straps

Steel wall bracing straps are 16-gauge steel 1½ inches wide. You can use them in place of a 1 × 4 wood let-in brace in stud walls. They keep the wall from racking during construction, but they aren't an effective component in a shear wall in strong wind storms or earthquakes.

T- and L-straps

Use T- or L-straps to connect wood beams to wood posts. Bolt the horizontal portion to the beam, and the vertical portion to the post. A T- or L-strap will give some lateral strength to a connection when you install it correctly. Connect T- or L-straps to the wood with $^3/_8$- to ¾-inch-diameter bolts or 16d nails.

H-Clips (Ply Clips)

Use H-clips to anchor the unsupported edges of plywood sheathing. Insert the clips between the panels along the unsupported edges of the plywood. This eliminates the need for wood blocking except where maximum shear strength is required. Some clips are made of extruded aluminum alloy and are tapered at the edges to make them easier to install. These clips are made for $^3/_8$-, ½-, $^5/_8$-, ¾-, and $^{13}/_{16}$-inch plywood.

Metal Bridging (Steel Cross-Bridging)

You can use metal bridging instead of wood cross-bridging or solid blocking in roof or floor framing. The cross-bridging prevents joists and rafters from warping and buckling and distributes loads throughout the frame. Metal bridging units are made of 16-gauge cold-rolled steel. The bridging is made for 2 × 8 to 2 × 16 members spaced on 12-, 16-, and 24-inch centers.

You have to nail some types of metal bridging while other types have spiked projections at the ends that you can nail into a wood member. Use 9-gauge nails 1½ inches long to fasten metal bridging to wood members.

Retrofitting

Since the recent hurricanes in Florida and earthquakes in California, many special types of retrofit lumber hardware have been designed to make existing buildings more resistant to lateral and uplift forces. Use this special retrofit hardware to:

▮ Attach the sill plate to the foundation

▮ Brace the underfloor framing to the foundation

▮ Transfer vertical loads from shear wall to shear wall

▮ Tie posts to the foundation

▮ Anchor floor joists to the foundation

▮ Anchor floor joists to the mud sill

▮ Tie floor joists to the foundation wall

▮ Tie studs to the sill plate

Figures 4-26 and 4-27 are before and after views that show some typical retrofits. Also look at Figure 2-13 to see post and wall configurations.

When you retrofit a damaged wood frame building, check with the building department to determine which code you must follow. You might be allowed to follow the code that was in effect when the building was originally built. Or you might be required to follow the latest code. The City of Los Angeles, newly-concerned about earthquake damage, has provided the following guidelines regarding their code:

1) When the damage to the lateral resisting elements or shear walls (in a single floor) along a particular shear line represents less than 10 percent of the total shear-carrying capacity of that shear line, the damaged section(s) may be replaced with the same construction as before.

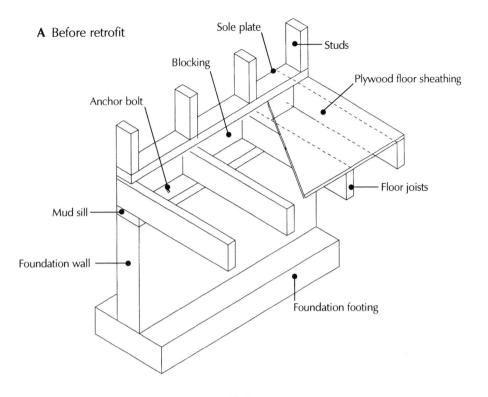

A Before retrofit

Sole plate

Studs

Blocking

Plywood floor sheathing

Anchor bolt

Mud sill

Floor joists

Foundation wall

Foundation footing

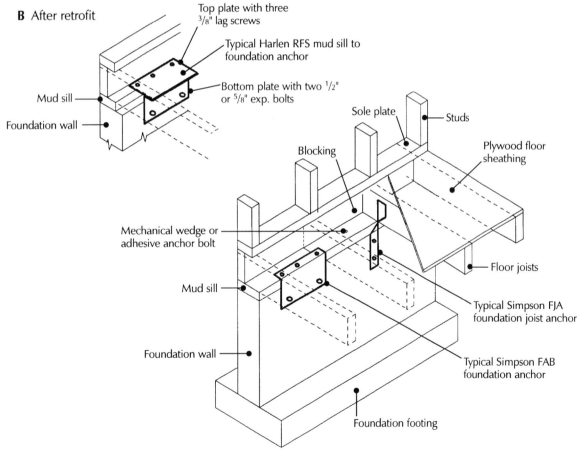

B After retrofit

Top plate with three ³/₈" lag screws

Typical Harlen RFS mud sill to foundation anchor

Bottom plate with two ¹/₂" or ⁵/₈" exp. bolts

Mud sill

Foundation wall

Sole plate

Studs

Plywood floor sheathing

Blocking

Mechanical wedge or adhesive anchor bolt

Floor joists

Mud sill

Typical Simpson FJA foundation joist anchor

Foundation wall

Typical Simpson FAB foundation anchor

Foundation footing

Figure 4-26 *Foundation wall without cripples*

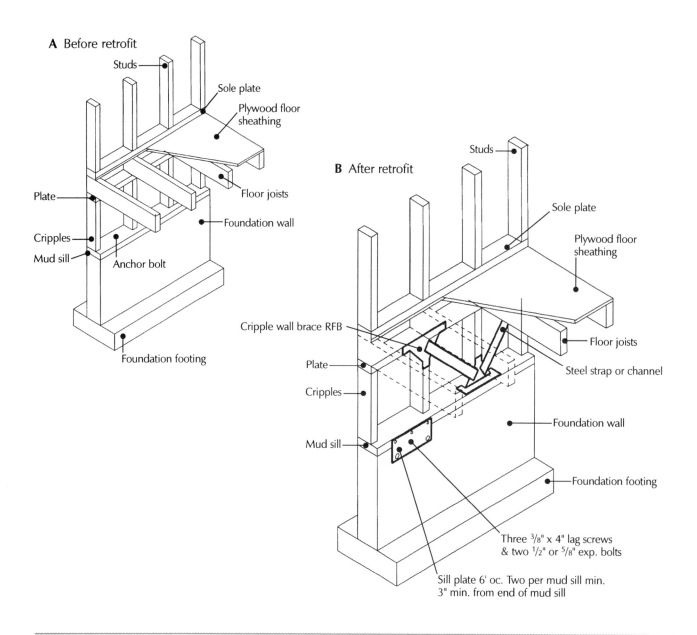

A Before retrofit

Studs

Sole plate

Plywood floor sheathing

Plate

Cripples

Mud sill

Anchor bolt

Floor joists

Foundation wall

Foundation footing

B After retrofit

Studs

Sole plate

Plywood floor sheathing

Cripple wall brace RFB

Plate

Cripples

Mud sill

Floor joists

Steel strap or channel

Foundation wall

Foundation footing

Three ³/₈" x 4" lag screws & two ¹/₂" or ⁵/₈" exp. bolts

Sill plate 6' oc. Two per mud sill min. 3" min. from end of mud sill

Figure 4-27 Foundation wall with cripples

2) When the damage to the lateral resisting elements (in a single floor) along a particular shear line represents 10 percent or more of the total shear-carrying capacity of that shear line, all damaged and undamaged lateral resisting elements along that shear line shall be redesigned to meet current code requirements.

3) When the total damage to all lateral resisting elements (in a single floor) in the same (east-west or north-south) direction represents 50 percent or more of the total shear-carrying capacity of that floor (in that direction), the entire floor's lateral resisting system in that direction shall be redesigned to meet current code requirements.

4) When designing for the required lateral resisting elements along a particular shear line, stucco and drywall shall not be given any shear value when used in the same shear line as a plywood shear wall.

5) Any damaged floor area of a building that must first be demolished before it is rebuilt shall be designed per current code.

Defects and Failures in Connections

To troubleshoot defective wood framing, begin by searching for design errors, poor wood quality, and poor workmanship. You can use the checklist in Figure 4-28 as a guide to finding common timber framing connection problems.

Excessive Deflection

Excessive deflection is usually the result of a design error. Building codes and standard practice set the maximum deflection for beams and trusses, but this is only a guide. Under certain conditions, following just the minimum requirements of the code can get you in trouble.

Most tables for selecting floor joists use $1/360$, $1/240$, or $1/180$ of the span as the maximum allowable deflection. Using $1/360$ of the span for the maximum allowable deflection, a joist spanning 30 feet must be designed so that it will deflect (sag) only 1 inch at the center of the span ($30 \times 12/360 = 1$ inch). If a deflection of $1/180$ of the span is used in the design, the allowable deflection will be 2 inches. This may not be noticeable to the building occupants unless the joist is located next to a wall that doesn't deflect, causing an objectionable floor slope between the wall and the first joist.

The same thing happens when a long joist is installed next to a short joist. The deflection of a joist with a shorter span will be less than that of a longer span. If the two joists are close together, the midpoint of the longer joist will be lower than the midpoint of the shorter joist. Another problem with sizing a joist to maximum deflection is that the floor becomes bouncy and noisy.

Buckling and Notching

Buckling in a floor or roof system is caused by a lack of bridging or blocking between joists or rafters. Buckling in a wall is caused by the lack of lateral support, or by eccentric loading.

Notching the bottom of a beam to accommodate ductwork, piping, or conduit will weaken the beam. Take this into account in your calculations.

Hostile Environment

❐ Check exposed wood connections for corrosion, loosening, and weakening.

❐ Check to see if the wood has been treated to prevent deterioration by fungus or termites.

Improper Nailing

❐ Check the size and penetration of nails.

❐ Check the number of nails used at connections.

❐ Check the exposure to weather. Wet conditions corrode nails and weaken wood.

❐ Check nail spacing and distance of nails from the edges of members.

❐ Check nailing in the load path of a shear wall. This is shown in Figures 3-36, and 3-38 in Chapter 3.

Improper Bolting

❐ Check joints connected by bolts, drift pins, lag screws, or wood screws for exposure to weather.

❐ Check that bolt holes aren't oversized or undersized, causing loose connections.

❐ Check connections for a tight fit.

❐ Check for loose or missing nuts and washers.

❐ Check to see if bolts are too close together, causing the members to split. See Figure 4-13 for recommended spacing of bolts. Also, bolts may be too close to the edge of the end of the member, causing splits. Stagger bolts in two lines.

❐ Make sure there are enough bolts in each connection.

❐ Check the angle of the wood grain at each bolted connection. An excessive grain angle weakens the connection. Forces through bolted connections are strongest when the force is parallel to the wood grain, and least when perpendicular to the grain. The strength of a bolted connection is reduced according to the angle between the force and the grain.

Improper Connections

❐ Check that connectors follow what is shown on the plans.

❐ Check that metal connectors, nailing, and bolting have been approved by the building department. Following an earthquake or severe wind storm, check whether the building department has lowered the approved values of metal connections. Allowable holding capacity of hold-down anchors, nails, and bolts were reduced by 75% following the Northridge, California earthquake.

❐ Make sure that members have full bearing at the seats of connectors. See Figure 4-22.

❐ Check for wood shrinkage.

Figure 4-28 *Checklist for troubleshooting wood framing*

Chapter 5

Concrete Formwork

Just because concrete formwork is temporary doesn't mean it can be done carelessly. You've got to design, engineer and build it accurately — the shape, position and finish of the concrete depend on it. It also has to withstand high pressures with little deflection, and all dead and live loads without collapsing. Finally, you need to build the formwork efficiently and economically, without sacrificing quality or safety.

Form Design

Concrete details in the plans and specifications should show the finished shape of the concrete and the texture of the exterior concrete surface. On major jobs, engineers have to design and approve the forms and shoring, but on most jobs, carpenters design and build the formwork.

Each part of a forming system has a specific function. When they work together, the entire system is stable, watertight, and resistant to the hydrostatic pressure exerted by the wet concrete.

Concrete Pressure

When you place fresh concrete in a form to make a wall, it's in a semiliquid state that presses out against the forms. This pressure is called *hydrostatic*, since it exerts the same force in all directions. Your job is to design and build forms that can resist this pressure. Note that the pressure isn't affected by the thickness of the wall. The pressure at any location up the wall depends on several factors, but surprisingly, the

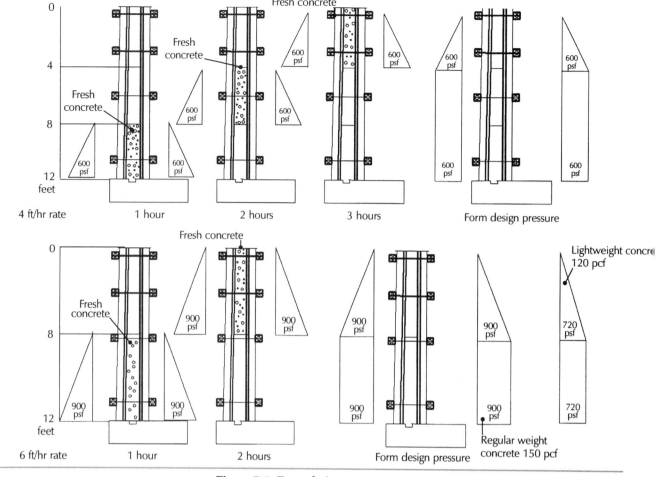

Figure 5-1 *Form design pressures*

thickness of the wall isn't one of them. The lateral pressure at any location on the wall is affected by:

1) *The height of the liquid concrete*: You can find the pressure at any point on the forms by multiplying the weight of concrete by the distance from that point to the top of the liquid concrete.

2) *The rate of pour*: The maximum pressure occurs at the bottom of the forms and decreases toward the top of the forms. Look at the hypotenuse (inclined leg) of the right triangle in Figure 5-1. A fast rate of pour results in forms supporting all liquid concrete, whereas a slow pour rate allows the concrete at the bottom of the forms to set up and become self-supporting before liquid concrete reaches the top of the forms. Figure 5-2 shows how the lateral pressure of concrete on forms increases as the rate of pour is increased. For example, at 70° F, concrete placed at a rate of 5 feet per hour exerts a maximum pressure of about 840 psf on the lower portion of the lift, while concrete placed at a pour rate of 7 feet per hour exerts about 1100 psf.

3) *The weight of the concrete*: Regular liquid concrete weighs between 150 and 160 pounds per cubic foot. Lightweight concrete weighs between 90 and 115 pcf, depending on the aggregate used.

4) *The surrounding temperature*: Concrete sets up faster at higher temperatures, so, since more of the concrete is self-supporting, the pressure is reduced. Figure 5-2 shows how concrete pressure varies with different pour rates and ambient temperatures. For example, concrete placed at the rate of 6 feet per hour exerts a maximum pressure of about 980 psf at the lower portion of the lift at 70° F. and approximately 1980 psf at 30° F.

5) *The type of cement (I, II, III, IV or V)*: Type I normal portland cement is the common type of cement used in construction. Use Type II modified portland cement for massive concrete jobs. It's moderately resistant to sulfate action. Use Type III high-early-strength cement when you need concrete to cure rapidly — perhaps on a driveway where early access is important. Type IV low-heat portland cement cures slowly, so you can use it also for massive jobs. Use Type V sulfate-resistant portland cement for concrete exposed to severe sulfate action, such as sea water or acids.

6) *The intensity of the vibration of the mix*: Vibration helps consolidate concrete, but it also slows the setup process.

7) *The concrete slump*: Slump depends mainly on how much water you put in the mix. Water produces a heavier mix and slows the setup process.

8) *The types of chemical additives in the concrete*: Some chemical additives, such as calcium chloride, speed up the setup process.

To find the pressure at any point on a form, multiply the weight of concrete by the distance from that point to the top of the liquid concrete.

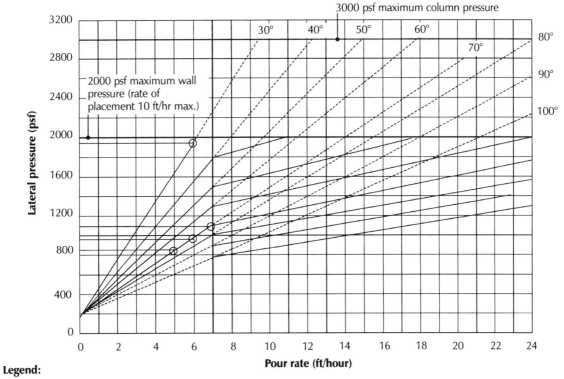

Figure 5-2 *Lateral concrete pressure at various temperatures*

Common Causes of Formwork Defects and Failures

Some common causes of defects or failure in concrete formwork are:

- Not enough diagonal and lateral shore bracing
- Placing concrete too quickly
- Unstable or frozen soil beneath mud sills supporting the shores
- Insufficient nailing
- Shoring not plumb
- Unsecured locking devices on metal shoring
- Vibration from concrete vibrators or moving loads such as concrete buggies
- Removing supports or forms too soon
- Knotholes, cracks, or other blemishes from the sheathing surface
- Careless reshoring
- Inadequate size and spacing of shores or reshores
- Faulty formwork design

The purpose of this chapter is to help you avoid making these costly and potentially dangerous mistakes in your formwork.

Wall Forms

There are a variety of foundation walls that require forms. Figure 5-3 shows typical foundations and footings for one- and two-story buildings. Use this figure only as a guide for footing types, since there are many things you need to consider about which size footing you should use. For example, you have to find out the depth of the frost line in the area where you're building. Foundations and footings must extend below the frost line. Check with your local building department to find out the depth of the frost line. In some portions of the northern United States, the frost line is 72 inches below grade.

A wall form may use sheathing, studs, wales, braces, stakes, shoe plates, spreaders, and form ties. Snap ties are used more often than form ties because they're easier to install. Figure 5-4 shows typical parts of formwork for a grade beam and Figure 5-5 shows one method of making a form for a concrete wall.

Sheathing

The sheathing you use in a form is the mold for the outside surfaces of concrete, so make sure it's smooth, tight, and strong enough to withstand the pressure of fresh concrete.

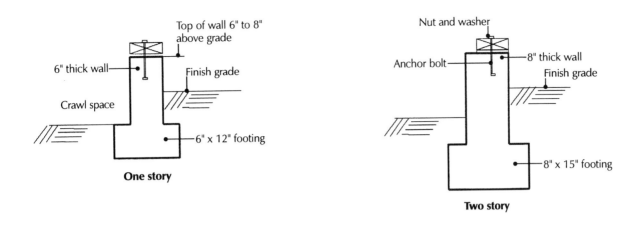

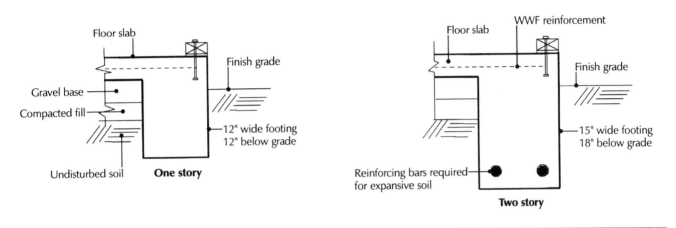

Figure 5-3 Typical dwelling foundation

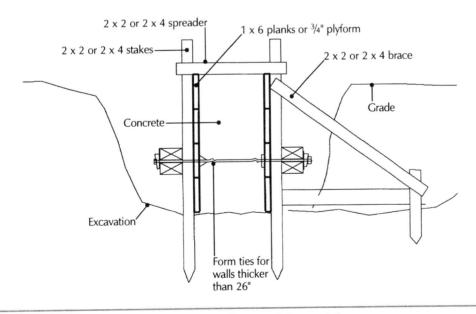

Figure 5-4 Formwork for grade beam

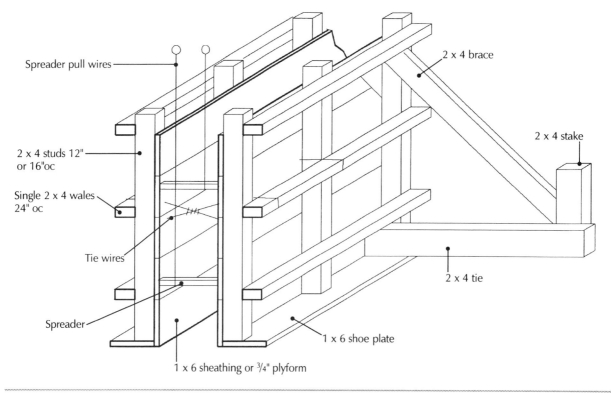

Labels in figure:
Spreader pull wires
2 x 4 brace
2 x 4 stake
2 x 4 studs 12" or 16"oc
Single 2 x 4 wales 24" oc
Tie wires
2 x 4 tie
Spreader
1 x 6 shoe plate
1 x 6 sheathing or ³⁄₄" plyform

Figure 5-5 *Formwork for a wall*

There are many materials you can use for building forms, including wood, plywood, steel, plastic, fiberboard, and certain types of composite wood board. Use wood that's straight, smooth, and kiln-dried. Extremely dry wood with a moisture content of less than 19 percent may swell when it comes in contact with wet concrete.

Plywood

If you use plywood for the sheathing, choose exterior-grade ¹⁄₂- to 1-inch plywood form sheathing with sanded surfaces. Always install plywood with the face grain perpendicular to the supporting members. Plyform is mill-oiled and you can use it many times.

One popular type of sheathing is plywood overlaid with a plastic coating. This material is called high-density overlaid (HDO) plywood. It has an overlay material made with a semi-opaque phenolic resin-impregnated fiber surface applied to one or both sides of an exterior-type plywood core. HDO plywood is expensive, but it gives a smooth concrete surface and you can use it many times. Other types of plastic-clad plywood include medium-density overlay and Plyron. Be sure that any reconstituted wood core board you select can be used with wet concrete.

Wood Boards

You can build field-fabricated forms with wood boards. You usually use wood boards for forming the edges of slabs and for constructing complex forms for beams and girders. You can use boards 1 or 2 inches thick, depending on the spacing between supporting studs and the lateral pressure of the wet concrete. Wood board sheathing isn't widely used anymore because it takes too long to install and it leaves an unattractive concrete surface finish after the forms are stripped. For a smooth surface, use T&G boards.

Fiber Reinforced Plastic (FRP)

You can also use fiber reinforced plastic (FRP). This has a smooth surface and it's easy to strip. Plastic form liners make a smooth or textured concrete surface. You can get liners that have vertically ribbed or fluted surfaces which give a board or brick appearance to concrete. Other types of FRP forms are used to replace board forms for one- and two-story buildings. Custom-designed column forms such as hexagonal-shaped columns are often made of FRP. To cast a round column, use a tubular fiber form made of asphalt-impregnated cardboard.

When you want a patterned architectural concrete finish, use a sheathing whose inside surface has the texture you want. You can do this by attaching a textured plastic liner to the inside of the forms. If specifications call for an aggregate finish, bond pea gravel to the sheathing with an adhesive such as epoxy.

Composite Wood Panels

Some of the composite wood panels can be used for load-carrying formwork sheathing provided they're backed with 1 × 4s or 2 × 4s spaced 6 inches on center. The weaker panels can be used as liners to provide a smooth surface to the concrete.

It's important to select composite wood panels that are specially manufactured to be used in formwork. The edges must be sealed and the surfaces durable when exposed to wet concrete.

Hardboard is a fibrous-felted board made of wood fibers impregnated with drying oils, fused together with heat and pressure. Tempered hardboard impregnated with a type of drying oil may be used as a form liner or facing material. You can use hardboard wall liners backed up with 1 × 4s at 6 inches on center or hardboard slab forms backed up by 2 × 4s spaced 6 inches on center.

Particleboard is made of wood particles and wood fibers bonded together with synthetic resins or other bonding material. Certain types of particleboard have limited applications in formwork, but may be suitable as liners under some conditions.

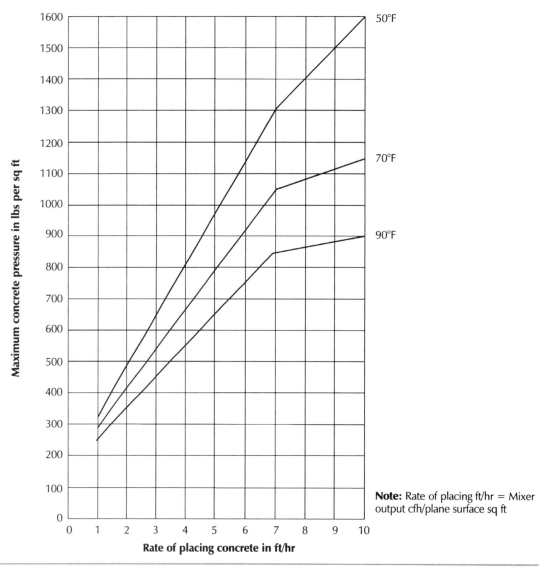

Figure 5-6 *Maximum concrete pressure*

OSB (oriented strand board) is made of flat wood chips or flakes all oriented in one direction and bonded with phenolic resins. There's a polymer overlaid OSB that you can use in formwork, but always check the manufacturer's specifications.

Form Studs

Form studs support sheathing against hydrostatic pressure of liquid concrete pushing the sheathing outward. Form studs are usually 2 × 4s or 2 × 6s spaced 12, 16, 24, or 32 inches on center. The usual spacing is 24 inches, but the maximum spacing depends on the type of sheathing and the lateral pressure of the concrete. Stud spacing depends on the type and thickness of the form sheathing, stud size, the maximum height of pour, and the ambient temperature. If three of these variables are selected, then adjust the other. See Figures 5-1, 5-2, 5-6, and 5-7.

For example, assume a maximum pressure of 600 psf and a 2-foot stud spacing. The pressure on each stud is 600 × 2, or 1200 plf. The span of the studs is determined by the spacing of the wales. Figure 5-7B shows that at 1200 plf, the maximum spacing of the wales and ties is 20 inches. If you use 4 × 4 studs, you can space the wales 31 inches on center. The American Plywood Association Publication V345 contains valuable information regarding concrete form design and engineering data.

You can prefabricate form panels by nailing the Plyform to the studs and plates. Prefabricated panels are normally 2 × 4, 2 × 8, or 4 × 8. Prefabricated panels save time since you only need to install the wales, braces, and ties to complete the formwork.

A Recommended maximum pressures on Plyform Class I (psf)

Support spacing (in)	Plywood thickness					
	15/32	1/2	19/32	5/8	23/32	3/4
4	2715	2945	3110	3270	4010	4110
8	885	970	1195	1260	1540	1580
12	335	405	540	575	695	730
16	150	175	245	265	345	370
20	—	100	145	160	210	225
24	—	—	—	—	110	120

Face grain across supports, deflection limited to 1/360th of the span and plywood continuous across two or more supports.

B Maximum spans for lumber framing (in)

Uniform load (plf)	Continuous over 2 or 3 supports (1 or 2 spans)							
	2 × 4	2 × 6	2 × 8	2 × 10	2 × 12	4 × 4	4 × 6	4 × 8
200	49	72	91	111	129	68	101	123
400	35	51	64	79	91	53	78	102
600	28	41	53	64	74	43	63	84
800	25	36	45	56	64	38	55	72
1000	22	32	41	50	58	34	49	65
1200	20	29	37	45	53	31	45	59
1400	19	27	34	42	49	28	41	55
1600	17	25	32	39	46	27	39	51
1800	16	24	30	37	43	25	37	48
2000	16	23	29	35	41	24	35	46

Spans based on single member allowable stress multiplied by a 1.25 duration of load factor for 7-day load. Deflection limited to 1/360th of the span with 1/4″ maximum. Spans are center-to-center of supports. Douglas-fir No. 2 or Southern pine No. 2

Figure 5-7 Maximum pressures on Plyform and spans for lumber framing

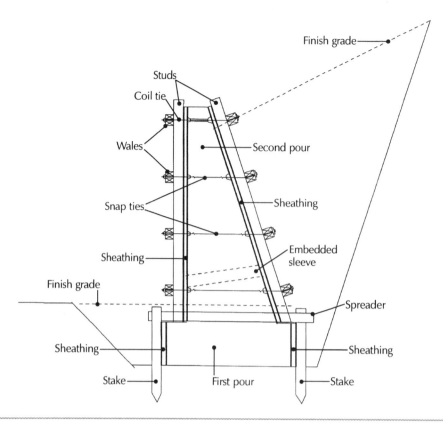

Figure 5-8 *Forms for a gravity retaining wall*

Shoe Plates

Shoe plates (or sole plates) help level the bottom of forms and provide anchorage for the bottoms of studs. Toenail or end-nail the studs to the shoe plate. See Figure 5-5. A shoe plate is usually a 1 × 6, 2 × 4, or 2 × 6.

Wales

You can support walls less than 4 feet high with stakes and braces as shown in Figure 5-4. Support the studs for walls over 4 feet high with wales (walers) as shown in Figures 5-8 and 5-9. Wales are usually single or double 2 × 4 lumber for light loads, and 2 × 6s for heavy loads. The wale spacing you use will depend on the strength of the studs and the lateral pressure of the concrete.

Form Braces

Bracing helps keep forms aligned and rigid. A typical brace is a 2 × 4 installed diagonally from a wale or stud near the top of a form to a stake driven into the ground. For added stability, install a horizontal member from the stake to a stud or shoe plate. See Figure 5-5.

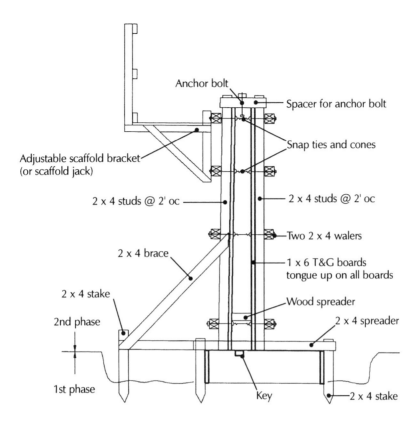

Figure 5-9 *Forms for a concrete retaining wall*

Stakes

Stakes are usually 2 × 4s driven into the ground. You can also use pointed steel stakes.

Spreaders

Spreaders are removable pieces of wood you set between sheathing to hold it apart. Spreaders are held in place by friction but you can pull them out through the wet concrete with an attached wire. See Figure 5-5. If you use snap ties, you don't need spreaders.

Form Ties

Tie wire is usually made of 8- or 9-gauge soft black annealed iron wire. Attach tie wires to each stud at the wales on both sides of a form. If you find that the wire isn't strong enough to carry the load, leave the studs in their original positions and space the wales closer together. Tighten the wires by twisting them with a wedge to prevent the forms from being pushed apart by the outward pressure of the concrete. See Figures 5-5, 5-10 and 5-11.

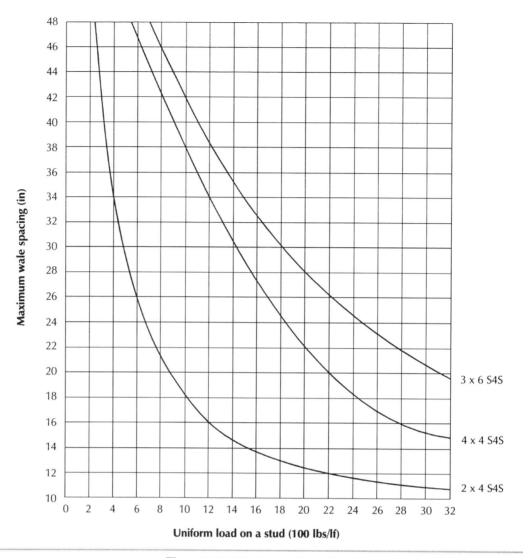

Figure 5-10 *Maximum tie spacing*

The approximate breaking strength of double-strand steel wires is:

- 8 ga 1700 lbs
- 9 ga 1420 lbs
- 10 ga 1170 lbs
- 11 ga 930 lbs

Here's the formula you can use to figure out the maximum tie wire spacing:

Maximum spacing (in) = Wire strength × 12 ÷ Uniform load on wales

For example, if the maximum pressure is 600 psf and wales are 2 feet on center, the uniform load on each wale is 600 × 2 = 1200 plf. Since 9-gauge wires have a breaking strength of 1420 pounds, the maximum spacing of the wires is 1420 × 12 ÷ 1200, or 14 inches.

To install twisted wire ties:

▮ Bore small holes through the sheathing at each side of the stud where you want the wire tie.

▮ Place the wires through both sides of the forms and around the wales.

▮ Pull the wires tight and twist the ends together.

▮ Place wood spreaders between the forms and twist the tie wires with a stick until the form walls are tight against the spreaders.

There are many types of form ties that can be broken within the concrete surface after you strip the forms. Some of these ties include:

▮ Snap ties with small cone spreader

▮ Washer spreader, crimped with break back

▮ Cone spreader

▮ Taper tie

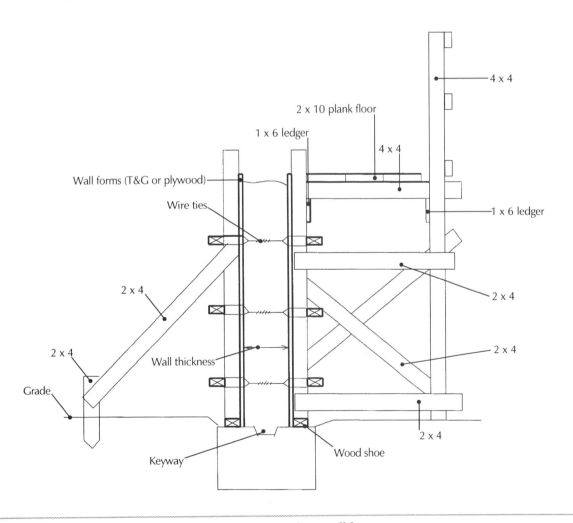

Figure 5-11 *Runway along wall form*

- Strap tie

- Loop end tie used with panels

- Straight tie with attached plastic tube

- Threaded bar with unattached plastic sleeve

- Coil tie with or without cone spreader

- Crimped tie with disconnecting ends

- Plain tie with shebolt disconnecting ends

Snap Ties

The main purpose of the form tie is to prevent the sides of a wall form from spreading apart. Many builders only use twisted wire ties on the simplest structures where strength of tie and appearance aren't important. Where appearance is important, there should be no exposed metal on the exterior concrete surface. Some job specifications require that all embedded metal be at least $1^1/2$ inches from the concrete surface. You can do this by using nonmetallic ties, such as fiberglass, or by using metal snap ties.

Snap tie is a general term that describes any tie that you can break off at a specified depth. After you snap the tie by twisting the end with a wrench, you should patch the tie end with a waterproof material that doesn't shrink, and that bonds well with the concrete. If any moisture reaches a metal tie end, rust stains will soon appear on the concrete surface. Use nonshrink patching grout or dry-pack mortar.

Some commercially available internal disconnecting ties, or snap ties, include:

- Snap ties with a cone spreader

- Crimped snap tie with disconnecting ends

- Plain tie with shebolt disconnecting ends

- Coil-type tie with or without a cone spreader

- Taper tie and bracket

Here are some safety rules for using snap ties:

1) Don't use bent snap ties.

2) Don't climb on installed snap ties.

3) Don't allow snap ties to remain in concrete more than 24 hours after placement. Remove at the break point as soon as possible.

4) Don't weld ties to any object.

5) Don't exceed safe work loads.

Form Hardware

Form hardware includes nails, screws, bolts, inserts, and sleeves. Use common nails to assemble form panels and other components that won't be disassembled when you strip the forms. Use 6d nails for 1-inch sheathing or ⁵⁄₈-inch plywood, and 3d blue shingle nails for fiberboard and thin plywood liners. Use double-headed nails to install kickers, blocks, braces, wales, or any other item that you have to remove when you strip the forms. Use bolts and lag screws for heavy formwork requiring 2-inch sheathing supported with heavy timbers.

Inserts

When working with concrete, you'll often have to use metal inserts to fasten beams and supported items such as pipe and conduit clamps, lifting lugs, anchors, and threaded inserts. Some inserts accommodate threaded bolts while others have a slot you insert the bolt head into. Sometimes you must use embedded weld plates so you can weld a steel member to the concrete.

Sleeves

Set sleeves in forms to accommodate piping, conduit, and ductwork. Then you won't have to core through hardened concrete and run the risk of cutting steel reinforcing bars. Make sleeves from galvanized steel sheets, plastic or cardboard. Use wood to form square or rectangular sleeve openings.

Sample Wall Form Design

Let's go through a comprehensive example for the design of a typical wall form. Figure 5-12 shows part of a typical wall form consisting of sheathing, studs, wales, ties and braces. The wall is 8¹⁄₂ feet high and 10 inches thick. The concrete pour rate is assumed to be 4 feet per hour. According to Figure 5-1, the maximum concrete pressure is 600 pounds per square foot (psf).

Assume that the studs will be spaced 1 foot apart to support the plywood sheathing every 12 inches. According to Figure 5-7A, Plyform Class 1 panels with a 600 psf load can be either ²³⁄₃₂ or ³⁄₄ inch thick. The face grain of the plywood can be either horizontal or perpendicular to the studs. We'll use ²³⁄₃₂-inch thick Plyform Class 1 panels.

Each stud must resist a uniform horizontal load of 600 pounds per linear foot since they're spaced 1 foot apart. The studs are supported laterally by the wales. Let's see if we can space the wales at about 48 inches apart. According to Figure 5-7B, 2 × 6 studs resisting a load of 600 plf shouldn't span over 41 inches. So we'll use 2 × 6 studs at 12 inches on center supported by wales spaced 41 inches on center.

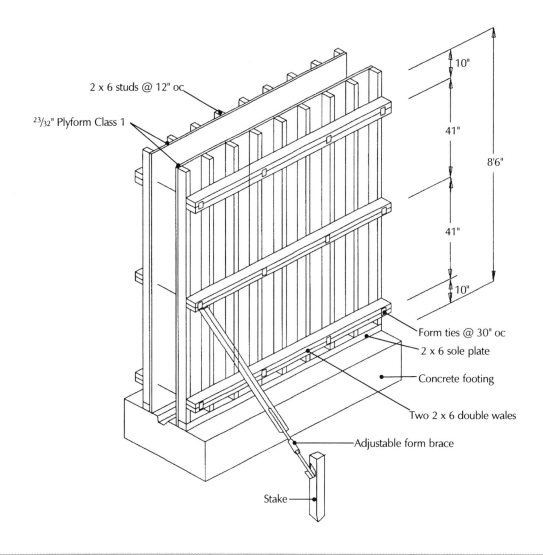

2 x 6 studs @ 12" oc

²³⁄₃₂" Plyform Class 1

10"

41"

41"

10"

8'6"

Form ties @ 30" oc

2 x 6 sole plate

Concrete footing

Two 2 x 6 double wales

Adjustable form brace

Stake

Figure 5-12 *Framing selection for a wall form*

The concrete pressure on the wales is calculated as follows:

Pressure = 600 psf × 41 ÷ 12 ft = 2050 plf

Since we'll use double wales, each wale carries 2050/2 or 1025 plf. The wale span is controlled by the spacing of the form ties.

Figure 5-7B shows that 2 × 6 lumber spanning 32 inches can support a load of 1000 plf. Since this is less than the 1025 plf load, set the form ties at 30 inches on center. The wales will be made of doubled 2 × 6 lumber.

Now we have to select the form ties. The tensile load in each tie is the load on the double wale times the spacing in feet:

2050 plf × 30 ÷ 12 ft = 5125 lbs

Wire ties won't work because an 8-gauge wire is only good for 1700 pounds. We must either reduce the spacing of the wire ties or select a manufactured form tie. See Figure 5-13 for some typical form ties. According to their manufacturers, here are some of the capacities of snap ties:

■ A $^9/_{16}$" shebolt has a working load of 6500 lbs.

■ A $^3/_4$" taper tie is good for 10,000 lbs.

■ A $^1/_2$" coil tie has a safe work load of 6750 lbs.

Let's use $^9/_{16}$" shebolt ties on 30-inch centers.

The wall forms should be braced for wind from either direction. Check with your local building department for the wind load in your area. A common wind load is 20 psf. Inclined bracing on one side of the forms is usually adequate for wind in either direction. To resist the wind uplift of the wall form, make sure the studs are nailed to the sole plate and the sole plate is anchored to the concrete footing. Attach the sheathing to the studs with as few nails as possible to make it easier to disassemble the forms. Gang forms may require additional nailing. Use 6d nails for $^{23}/_{32}$- and $^3/_4$-inch plywood. Don't butt the panels too tightly together as plywood has a tendency to swell when exposed to the wet concrete for the first time.

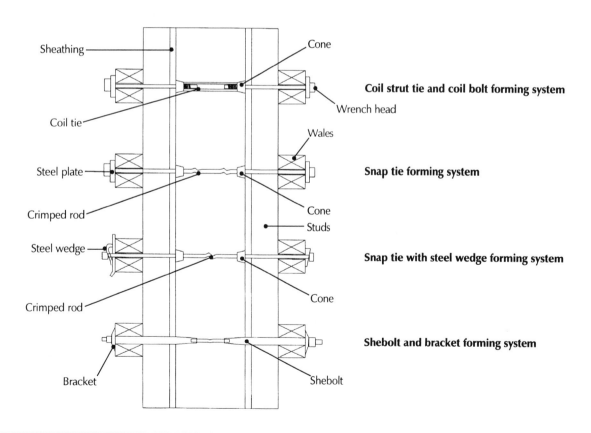

Figure 5-13 *Types of form ties*

Here are some rules that you should follow when designing wood forms:

1) Use stock sizes and lengths of lumber.

2) Use as few lengths of lumber as practical.

3) Use as few units as possible. You don't want to units to be too heavy, however.

4) Design the forms for easy stripping.

5) Provide for the reuse of such units as wall panels, floor panels, and beam and column forms.

6) Provide bevel cuts and keys so that the forms can be released with little prying.

Figure 5-14 is a checklist you can use when you're designing concrete wall forms.

❏ Determine what form materials are readily available.

❏ Determine the rate of delivery of concrete to the job site in cubic feet per hour.

❏ Determine the square feet of floor area the concrete will enclose.

❏ Determine the rate of pour in the forms in vertical feet per hour by dividing the rate of concrete delivery (cf/hr) by the floor area of the concrete (sq ft).

❏ Try to forecast the air temperature at the time you place the concrete. During cold spells on some jobs, you have to use oil-fired heaters to artificially heat forms and enclose the work with tarps.

❏ Use Figure 5-2 to find the maximum concrete pressure in the forms.

❏ Determine the maximum stud spacing.

❏ Determine the uniform load on each stud in plf.

❏ Determine the maximum wale spacing. Wale spacing depends on the strength of the studs and the lateral pressure of the concrete.

❏ Determine the uniform load on the wales in plf.

❏ Determine the maximum tie spacing and tie strength required.

❏ Compare the maximum tie spacing with the maximum stud spacing. The strength of tie wires is usually adequate for any spacing of the studs or wales.

Figure 5-14 Checklist for designing wall forms

Recommended Tolerances When You Build Forms

Concrete forms may vary from plan dimensions due to errors in building the forms or because of settlement and shifting of the formwork. But variations are acceptable within limits. The American Concrete Institute (ACI) recommends the following tolerances in formwork:

- The variations in concrete plan dimensions should be within -$^1/_2$ and +2 inches (from $^1/_2$ inch less than, to 2 inches more than, the plan dimension).

- The misplacement of concrete members should be within 2 percent of the footing width, but not more than 2 inches.

- The reduction in thickness of members should be within 5 percent of the specified thickness.

- The variation in plumb for concrete walls shouldn't be more than ± 1 inch for structures up to 100 feet high.

- The variation in plumb for conspicuous lines such as control joints shouldn't be more than ± $^1/_2$ inch for walls up to 100 feet high.

- The variation in any wall opening dimension should be within -$^1/_4$ and +1 inch.

- The variation in wall thickness should be within:

  -$^1/_4$ and +$^3/_8$ inch for walls up to 12 inches thick
  -$^3/_8$ and +$^1/_2$ inch for walls 12 to 36 inches thick
  -$^3/_4$ and +1 inch for walls over 36 inches thick

- The elevation of a concrete soffit should be within ± $^3/_4$ inch of the specified elevation.

- The variation in an elevated slab's thickness should be within:

  -$^1/_4$ and +$^3/_8$ inch for slabs up to 12 inches thick
  -$^3/_8$ and +$^1/_2$ inch for slabs 12 to 36 inches thick
  -$^3/_4$ and +1 inch for slabs over 36 inches thick

- The variation in any slab opening dimension should be within -$^1/_4$ and +1 inch.

- The variation from level for a beam soffit should be within ±$^3/_4$ inch over the entire length.

- The variation from level for an exposed parapet should be within ±$^1/_2$ inch over the entire length.

- The deviation from any cross-section dimension should be within:

  -$^1/_4$ and +$^3/_8$ inch for thicknesses less than 12 inches
  -$^3/_8$ and +$^1/_2$ inch for thicknesses 12 to 36 inches
  -$^3/_4$ and + 1 inch for thicknesses over 36 inches

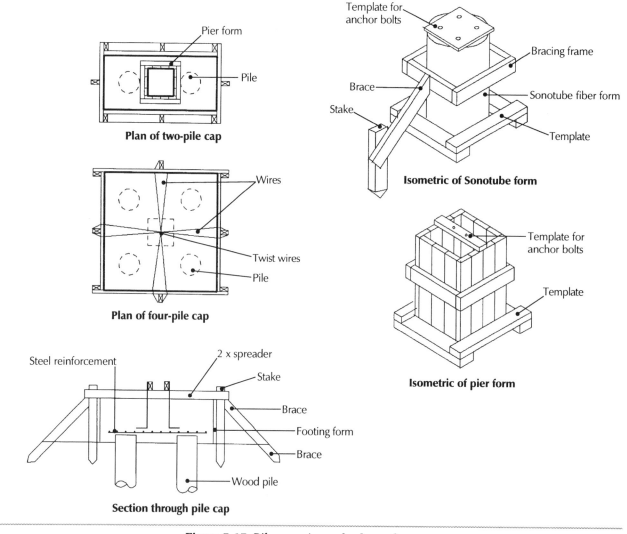

Figure 5-15 *Pile cap, pier and column forms*

Column Forms

Referring to Figure 5-15, the major components of column forms include:

▌ Templates set on the foundation for positioning the column forms

▌ Forms made of wood, prefabricated steel, a combination of steel and plywood, or fiber reinforced plastic (FRP)

▌ Bracing to stabilize and align the forms

There are some accessories you must have for column forms. Use chamfer strips to smooth out sharp corners of rectangular columns. Wood or steel yokes and locks hold the sheathing in place and resist the lateral pressure of the concrete. Space yokes vertically according to the largest dimension of the rectangular column cross section and the height of the column. The wider side of the column will have the longer yoke, which suffers the greater bending stress. The spacing may vary from 12 to 30 inches for columns 16 to 36 inches wide and up to 20 feet tall. Figure 5-16 shows the suggested spacing of column yokes. Here's how to use the table:

- Select the height of the column along the left side of the table.

- Select the largest cross section dimension of the rectangular column at the top of the table.

- Find the point those values intersect in the table to find the maximum spacing of the column yokes.

For example, if the column height is 12 feet and the largest cross section dimension is 18 inches, the maximum yoke spacing is 18 inches.

You can use steel column clamps instead of wood yokes. Since steel yokes require the same spacing as wood yokes, use Figure 5-16 to find the spacing required for various types of metal column clamps.

	Largest dimension of column (inches)							
Height (ft)	16	18	20	24	28	30	32	36
1		29	27		21	20	19	17
2								
3	31			23				
4			26		20	19	18	15
5		28						12
6					18	18	17	11
7	30			22	15		13	
8			24		13	12	12	10
9	29	26		16	12	10	10	8
10			19					
11	21	20	16	14	10	9	8	7
12		18		13	9	8		
13	20	16	15	12			7	6
14			14			7		
15	18	15	12	10	8	6	6	
16	15	13	11		7			
17	14			9	6			
18		12	10					
19	13			8				
20		11	9					

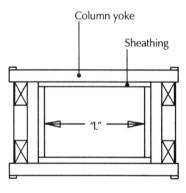

Figure 5-16 *Spacing of column form yokes*

Here's how you design and build a column form:

1) Determine what form materials are readily available.

2) Determine the height and cross-sectional dimensions of the column.

3) Determine the yoke spacing.

4) Build and install a template at the base of the column.

5) Build and erect the column forms and yokes.

6) Brace and align the column forms in both directions.

Column Anchor Bolts

Install wood templates to hold column anchor bolts in position in a slab or concrete wall. See Figure 5-17. Secure each bolt to the template with a pair of nuts, one below and one above the template. Carefully locate the template, using a steel tape and chalk line. You can remove the upper nut and template after the concrete has set up, leaving the projected end of the threaded bolt above the top surface of the concrete. You can install anchor bolts connected to hold-down hardware with a $1/8$-inch tolerance since the bolts are inserted into slightly oversized holes.

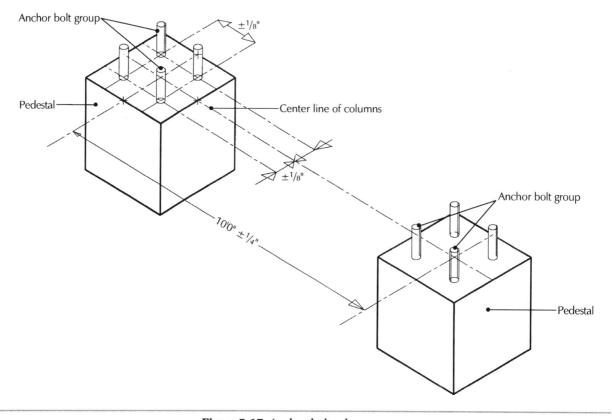

Figure 5-17 *Anchor bolt tolerance*

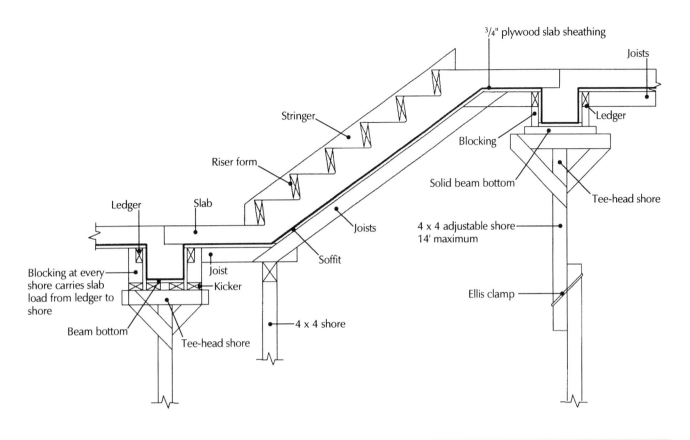

Figure 5-18 *Forms for concrete stairs*

Elevated Slab Forms

The major components you need to form an elevated concrete slab include shores, beams, stringers, and sheathing. See Figure 5-18. You can use 4 × 4 wood posts or tubular steel shores to support the beams. You can adjust the height of a wood shore with a turnbuckle assembly or Ellis clamps.

Steel shores have telescoping steel tubes with holes that you can align and insert pins through to vary the height of the shore from 6 to 14 feet. You can turn a screw at the top of the shore to fine-tune its final height. The manufacturers of various tubular shores provide tables showing the capacity of each type of shore at various heights. The longer the shore, the less it can support. For example, a wood shore 6 feet tall can support 7500 pounds, but a 14-foot shore can support only 2000 pounds.

Beams support stringers (joists) which support sheathing. These beams can be solid wood, laminated wood, box beams, or I-beams. Some beams are made of lightweight aluminum that you can adjust from 4 to 20 feet long.

You can use 1 × 6 tongue-and-groove boards or plywood sheathing to form the soffits of elevated slabs, or you can use ribbed or corrugated steel decking.

Here's how you erect forms for an elevated slab:

1) Erect 4 × 4 posts or steel shores. Install mud sills if the shoring is supported on earth.

2) Install beams over the shores and bracing between them. Brace steel shores with steel rods, angles, or channels.

3) Install stringers perpendicular to the beams.

4) Install sheathing over the stringers.

Design loads for slab forms are:

Slab thickness (in)	Design load (psf)
4	100
5	113
6	125
7	138
8	150
9	163
10	175

These values include a 50 psf load for workers, non-motorized equipment, and impact. They also include the weights of beam forms, stringers, and sheathing. If you use motorized buggies, add 25 psf to the loading.

Safety Rules for Shoring

Here are some safety rules you should follow when you install shoring:

▎ Follow local ordinances, codes, and regulations for shoring.

▎ Use the manufacturer's recommended safe working loads for the type of shoring frame and the height from the supporting sill to the formwork. To prevent accidents, the shore designer must consider the weight of wet concrete and forms, the strength of the stringers, joists, sheathing, and the capacity of the shores.

▎ Keep a shoring layout at the job site. A shoring layout is a drawing prepared by the job superintendent or the engineer that shows the location and size of shores, bracing, stringers, and joists. When you rent shoring equipment, the rental company usually prepares the shoring layout. Don't exceed the shore frame spacing or shore heights shown on the shoring layout.

▎ Provide mud sills to maintain a solid footing under each shore to distribute loads properly. Be sure the soil under the mud sill is stable.

▎ Use adjustment screws to adjust the shore heights over uneven grade conditions.

▎ Plumb and level all shoring and formwork before, during, and after you pour the concrete.

- Fasten all braces securely and watch them constantly for potential form failure, especially during a concrete pour.

- Don't remove braces or back off adjustment screws until the job superintendent or engineer tells you to.

- Have a reshoring procedure approved by a qualified engineer.

Reshoring an Elevated Slab

Reshoring is when you reinstall the supports of an elevated slab form after you've partially stripped the form. In some cases, where you've stripped the forms to reuse them, you'll have to reshore the concrete until it gets its full strength. Don't wedge reshores so tight that they lift the concrete and cause a reversal of stresses in the beams or slabs. This will cause cracking.

Curing Concrete

After you've poured concrete, you must allow it to cure properly. Concrete will continue to get stronger for years, but for all practical purposes, it gains most of its full strength (95 percent) in the first 28 days.

Curing continues as long as there's enough water to react with the cement. However, concrete that dries out too soon won't develop its full potential strength. For maximum strength, keep concrete moist during the entire 28-day curing period. In hot dry weather, cover it with wet burlap, straw, or sand. Or you can spray the concrete surface with a curing compound made of a blend of oil, resins, waxes, and solvents. When the solvent evaporates, a thin membrane will stay on the surface, sealing in the original mixing water.

Removing Formwork

Formwork around concrete must remain in place until the concrete is self-supporting. Don't strip the forms until concrete compressive strength tests show that the concrete has reached the minimum strength required to withstand all the anticipated loads.

The minimum time to wait before you can strip forms from concrete will depend on the type of cement you use in them. Type 3 high-early-strength portland cement cures much faster than Type I normal cement. When the engineer or the building codes don't provide specific stripping standards, use the American Concrete Institute (ACI) Committee 347 suggestions in Figure 5-19 as a guide.

Item	Curing time
Walls*	12 hours
Columns*	12 hours
Sides of beams and girders**	12 hours
One-way floor slabs	
Under 10-ft. span	4 days (design live load less than dead load) 3 days (design live load more than dead load)
10- to 20-ft. span	7 days (design live load less than dead load) 4 days (design live load more than dead load)
Over 20-ft. span	10 days (design live load less than dead load) 7 days (design live load more than dead load)

*Where forms also support formwork for slabs or beam soffits, use the removal time for the supported members.

**Where forms may be removed without disturbing shores.

Courtesy: American Concrete Institute

Figure 5-19 *Suggested curing times for concrete*

Maintaining Forms

Inspect plywood forms for wear each time you strip them. Maintain the forms by cleaning, repairing, priming, refinishing, and treating with a form-release agent each time you use them. Use a hardwood wedge and a stiff fiber brush to clean the forms.

You can also use acid washing for general cleaning. Wet the surface thoroughly with a 5 to 10 percent solution of muriatic acid, and scrub with a stiff bristle brush. Flush with clean water to remove the acid.

Remove all nails and fill holes with patching plaster or plastic wood filler material. You can also repair the forms by applying grout and rubbing it with burlap to completely fill all pits. To remove excess grout, rub with clean burlap. Then sand the forms with No. 2 sandpaper to remove all excess mortar and make the panels smooth and uniform in color and texture. Proper maintenance will increase the life of a form and consistently produce a smooth concrete finish.

Chapter 6

Building Layout

No form builder or framer should start any construction until he's *absolutely certain* the job site has been accurately surveyed and staked out. Construction work must be within the property lines. Violating a neighbor's property can cause serious legal problems. You also can't stockpile lumber on public property without permission. You'll get a citation for this. In short, you have to know where the property lines are before you begin work.

Occasionally, construction has taken place over property lines because someone assumed a surveyor's offset marker was the true property corner point. Surveyors often offset their markers when a true corner point is inaccessible, as shown in Figure 6-1. They also place reference, or witness, stakes near a corner monument as a safeguard against losing the main marker, as shown in Figure 6-2. A builder may inadvertently use one of the reference stakes as a corner point because the true corner monument was accidentally destroyed. This could cause him to build in the wrong place. Most land surveyors set a brass tag engraved with their R.C.E. or L.S. number, so if there's any doubt about a corner point, contact the surveyor. The layout person must interpret the surveyor's monuments correctly. He or she should compare the surveyor's plat with the architect's plot plan and the markers in the field. (A plat is a drawing showing the boundaries of an individual parcel of land.)

After finding the true property corners, the engineering surveyor or layout person can begin to set the building foundation lines. The layout person also sets bench marks within the construction site to control foundation elevations. It's important to have a common system of elevations with public streets and sewers so rainwater and sewage will drain properly from a site.

Engineering surveying is always associated with construction. Field surveying may be subcontracted to licensed land surveyors or done by the contractor's own layout personnel. An engineering surveyor doesn't have to be a licensed surveyor or

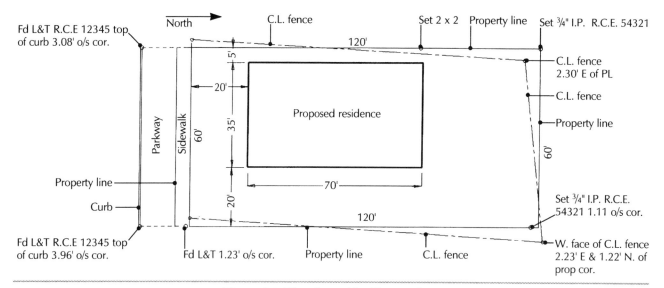

Figure 6-1 *Plot plan with offset markers*

registered civil engineer, but he or she should be qualified in the use of surveying instruments and related mathematics. A knowledge of trigonometry is essential. See Figure 6-3. An inexpensive pocket calculator that contains trigonometric functions helps.

The layout person controls all the elevations for a construction project from permanent monuments or bench marks. These may be a lead and tack in a concrete curb, the edge of a manhole cover ring, or a government monument (Figure 6-4). On large jobs, the engineering surveyor sets permanent bench marks at several locations on the site. On ordinary jobs, you can use 2 × 2 hardwood hubs.

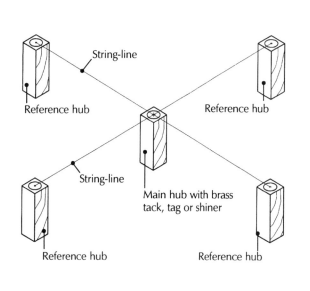

Figure 6-2 *Reference stakes*

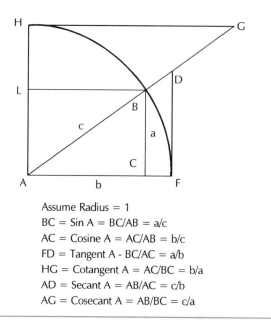

Assume Radius = 1
BC = Sin A = BC/AB = a/c
AC = Cosine A = AC/AB = b/c
FD = Tangent A - BC/AC = a/b
HG = Cotangent A = AC/BC = b/a
AD = Secant A = AB/AC = c/b
AG = Cosecant A = AB/BC = c/a

Figure 6-3 *Basic trigonometric formulas*

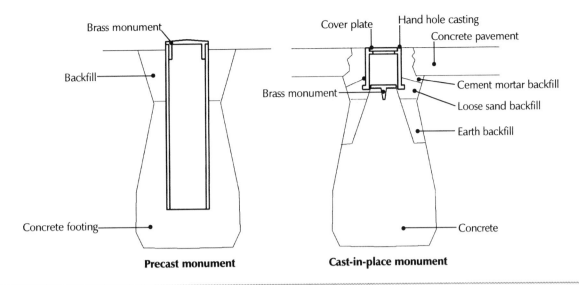

Figure 6-4 Typical concrete monuments

The field surveyor sets all transit lines, coordinate lines, and bench marks necessary for construction. These lines control the layout of a building and locate the datum elevation shown on the plot plan. Elevation control points may be the top of a street curb, or a manhole cover. A properly-drawn plot plan will show the location of all buildings, their setbacks, and side and rear yard lines. Figure 6-5 shows a plot plan with coordinate lines.

The surveyor should set reference stakes in case the main control marks are lost or damaged. He should flag these stakes with bright-colored strips of plastic to make them more visible. Finally, the layout person should prepare a neatly-drawn field book that reflects where all of the stakes are.

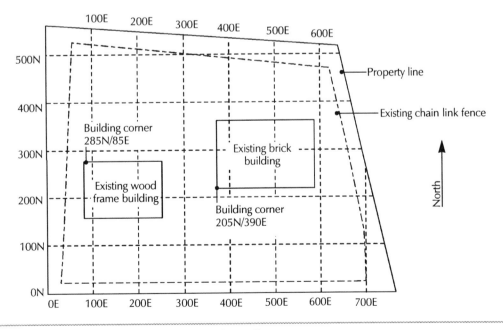

Figure 6-5 Typical coordinate system

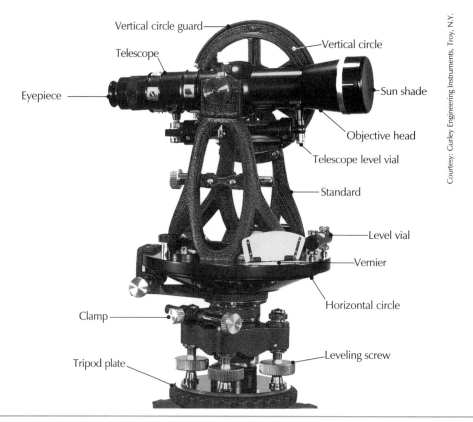

Figure 6-6 *Standard precise transit*

Survey Equipment

Transit

Surveyors use transits to locate horizontal and vertical angles. They also use them to set elevations, though they're not as effective as levels. A transit has more moving parts than a level and is more difficult to keep steady. Look at the standard precise transit in Figure 6-6.

A transit's components include a telescope, standards, plate, magnetic compass, horizontal and vertical circles, and verniers. The horizontal and vertical circles are divided into 30-minute increments. A vernier is a graduated scale used to measure subdivisions of the horizontal and vertical circular scales. You can read angles as small as 5 minutes on some built-in verniers. See Figure 6-7. The telescope of a transit has an eyepiece, cross hairs, focusing screws, bubble vials, and adjusting screws for leveling the instrument.

A vernier can be used in reading level rods, calipers and other precision instruments. The principle of a vernier is that the width of the divisions on the vernier is slightly less than the divisions on the scale. Figure 6-7 shows this principle.

The distance between zero and 10 on the vernier is the same as the distance between 3.00 and 3.90 on the scale. When zero on the vernier coincides with 3 on the scale, the reading is 3.00, as shown in section A of Figure 6-7. When 1 on the vernier coincides with 3.1 on the scale, the reading is 3.01, as shown in section B. When 8 on the vernier lines up with 3.8 on the scale, the reading is 3.08, as shown in section C.

A transit is mounted on a wood or aluminum tripod. A plumb bob, or plummet, is suspended under the transit over a selected point on the ground. To position a transit, set the tripod firmly on the ground with the telescope at eye level. Then adjust the leveling screws so the two bubble vials are horizontal. Rotate the plate and readjust the leveling screws until the plate remains level no matter which direction you point the telescope. The layout person using a transit needs to understand the following:

▌ Angles, bearings, and azimuths

▌ How to add and subtract bearings

▌ How to obtain angular differences

▌ How to read the verniers on the horizontal and vertical circles

Builder's transits are more rugged and less precise than the ones engineers use. A builder's transit normally has about half the magnifying power of an engineer's transit.

Most surveyors on larger jobs use electronic surveying equipment, like the automatic level in Figure 6-8. These instruments require fewer personnel to operate, and are far more precise.

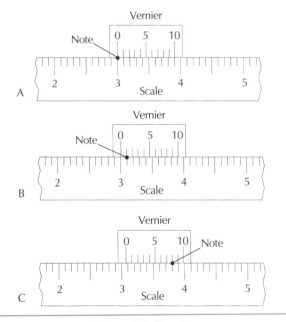

Figure 6-7 Verniers

Figure 6-8 Automatic level

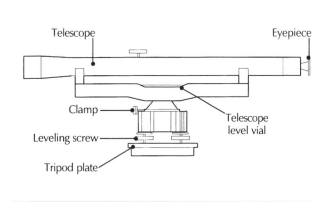

Figure 6-9 *Sketch of a Dumpy level*

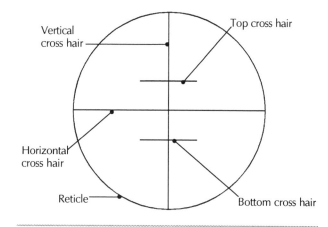

Figure 6-10 *Telescope cross hairs*

Levels

A basic level is similar to a transit except its telescope is longer, doesn't rotate vertically, and has greater magnifying power. Figure 6-9 is a sketch of a common type of instrument used in construction for decades. Most levels don't measure horizontal angles. A level is more practical than a transit in setting elevations because it has fewer moving parts and is easier to keep steady. Use a level to set the height of concrete forms and floor slabs.

The telescope of a basic level has an attached level vial. The telescope is mounted on the end of a straight bar. The bar rotates at its center on a vertical axis. By using the adjusting screws, the user makes sure the instrument is horizontal. When rotating the telescope, the operator is sighting on a horizontal plane. Any object that appears within the center horizontal cross hair of the telescope is at the same elevation as the center of the telescope.

In a telescope, there's a single vertical cross hair, one horizontal cross hair at the center, and two short horizontal stadia cross hairs. All the cross hairs are made of fine platinum wire. Figure 6-10 shows the typical arrangements of cross hairs in a transit or level. You use the stadia hairs to find a distance by reading the interval on a stadia or leveling rod. When stadia hairs read 1:100, it tells you that an increment of 1 foot between the upper and lower stadia hairs indicates the rod is 100 feet away.

You can read a typical builder's level to an accuracy of $^3/_{16}$ inch in a 150 foot distance. You can use a level for the following types of work:

■ Bench mark leveling

■ Transferring elevations from bench marks to work above or below ground (i.e., street and drainage works)

■ Profile leveling and cross-sectioning

■ Grading work

Caring for Your Transit and Level

Here are some suggestions on how to take care of a transit and level:

■ If it rains while you're surveying, cover the instrument with a plastic hood or any type of plastic bag.

■ Don't store a wet instrument in its metal carrying can. Let it dry indoors before storing it.

■ After surveying in cold weather, don't bring the instrument directly into a warm room. Don't expose the instrument to any sudden changes in temperature.

■ If the instrument fogs up, warm it with a heat lamp.

■ When you carry an instrument and tripod on your shoulder, tighten the clamps slightly to keep it from moving around. Always set it down gently to keep the cross hairs from breaking.

■ Make sure the needle on the magnetic compass is lifted when you carry the instrument.

■ Never tap a level with a hammer or other tool. This may damage the plaster of paris the vials are set in.

Tripods

Unless they're using heavy hardwood frames to mount electronic survey equipment, builders generally use lightweight tripods. The extendable legs of the tripod have metal shoes with spurs and hardened steel points. Use the spurs to help drive the shoes into the ground. Spread the tripod legs as you set and press them into the ground to keep the tripod from being pushed over. On slopes, place two legs of a tripod on the lower part of the slope and extend the third leg upward. See Figure 6-11.

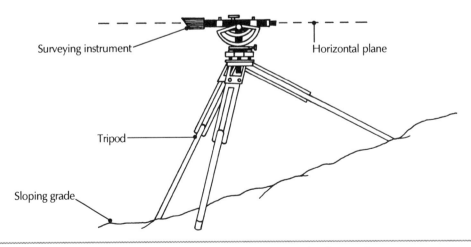

Figure 6-11 Setting up a tripod on a slope

Here's how to use a tripod:

■ Make certain the tripod is in good condition.

■ Check for looseness in the shoes, hinges, or where side pieces enter metal sockets.

■ Tighten all points with a tripod socket wrench.

■ Don't set a tripod on asphalt pavement during hot weather. The sharp points will sink into the pavement, causing the instrument to be off level.

■ Make sure the shoes of a tripod don't slip when you set it on concrete.

■ Rub a wood tripod regularly with linseed oil.

Plumb Bob or Plummet

This is a simple tool with a solid brass or steel cone weight suspended on a cord. The force of gravity pulls the cord in a vertical line. The cord is usually made of linen or nylon, in white or bright colors for easy visibility.

Handle a plummet carefully and take these steps to protect it:

■ Protect the point of the bob.

■ Avoid needless knots in the string.

■ When not in use, carry the plumb bob in its leather holster.

Leveling Rod

There are many types of rods used for vertical measurement. A builder's rod is usually 12 feet long, and folds into two sections. It's graduated in feet and inches to the nearest $1/8$ inch increment. An engineer's rod is marked in feet and decimal parts to the nearest $1/100$ of a foot. This rod is painted white, with feet marked in red and fractions in black. There's a round target painted black and red over a white background attached to the rod.

Another version of the rod is the measuring pole. This device extends to 26 feet in length and is self reading. You can measure height and depth using only a measuring pole.

Steel Tapes

Use a steel tape when you need to measure more than 6 feet. Tapes may be band chains, flat steel wire tapes, ordinary steel tapes and metallic tapes.

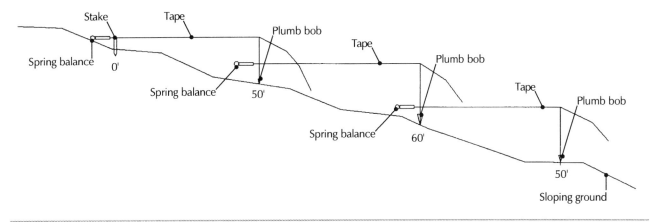

Figure 6-12 *Breaking tape on a slope*

Many old time surveyors still call a steel tape a "chain" because the early American surveyors used a chain made of metal links to measure distances. This led to the term "chaining" for measuring distances. Tapemen were called chainmen. Even today, some manufacturers of steel tapes call their product chain tapes.

Most steel tapes are 100, 200, or 300 feet long. They're marked in 1 foot intervals. One end is marked in tenths and hundredths of a foot, reading from right to left. Keep a 100-foot tape on a reel when you're not using it. Use leather thongs and holding clamps to keep the tape tight and tension handles to reduce sag. Another important accessory for the steel tape is a repair kit. This kit should have punch pliers, eye-letting pliers, eyelets, Allen wrench, pierced plain tape, dies, and rivets. Use this kit to repair a broken or kinked tape in the field to save time.

For level terrain chaining, lay the tape directly on the ground and mark points by steel pins or scratches on the sidewalk or pavement. In rough terrain or where there's vegetation, hold the tape horizontally above the ground. Attach a spring balance to one end of the tape to apply tension. Carry the points to the ground with plumb bobs suspended from a cord. Figure 6-12 shows how to measure horizontal distances along a slope.

A steel tape gives correct measurements when supported throughout its length at a temperature of 68 degrees F. and a tension of 10 to 12 pounds. If the same tension is used when the tape is suspended from both ends, the true horizontal distance between the ends will be shorter than what is shown on the tape. A light 100-foot tape, weighing 1 pound, would be shortened about 0.042 foot unless pulled with 15-pound tension. Figure 6-13 shows two chainmen measuring with a steel tape. Check the manual that came with the tape for the proper tension.

A tape can be held approximately horizontal with the aid of a hand level. Set temporary stakes in the ground along the line of measurement to help keep a straight line.

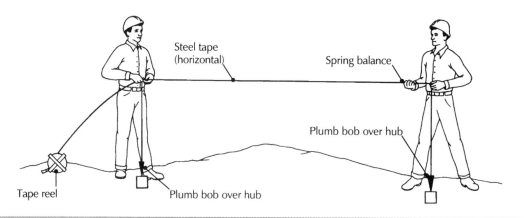

Figure 6-13 *Measuring horizontal distances*

The most common reading error in measuring horizontal distances is dropping 1 foot or 100 feet. Errors may be due to:

■ Incorrect reading of tape

■ Tape not held horizontal

■ Tape not straight

■ Incorrect tension

Steel tapes are easily damaged or made unreadable. Here are some precautions you can take to prevent this:

1) Be careful when any vehicular traffic has to cross a steel tape.

2) Don't jerk or step on a steel tape.

3) Avoid kinking a steel tape.

4) Coil band tapes in 5-foot loops, figure-of-eight form.

5) Wipe clean and dry an etched tape at the end of each day's work.

6) Use a tape repair kit if a chain is kinked or broken by a vehicle.

Spacing Rule and Tracing Tape

The spacing rule has been a favorite carpenter and surveyor tool for decades. It's a folding wooden rule usually 6 feet long when extended, and 6 inches long when folded. It's made of hardwood, with steel brass-plated spring joints that lock the rule in position. The surface has an enamel finish with black and red numbers printed on a white background. The carpenter's rule is graduated in feet and inches to the nearest $1/8$ inch. The surveyor's rule is marked in feet and tenths of a foot to the nearest $1/100$ of a foot.

A tracing tape is a 1-inch wide cotton tape, usually 200 feet long. Use it to lay out excavation or foundation lines.

Chalk Line and Straightedge

The chalk line is also called a snap line. It's made of a twisted cord, coated with white chalk, in a reel. To use a chalk line, stretch the line tight just above the surface between two points you want to connect with a straight line. Snap the cord to make a straight guideline.

A straightedge is made of a tapered wood or metal ruler with hand holes, 30 inches long at the bottom edge and 10 inches long at the top edge. Use this tool as a longer level with a small spirit level placed on the top edge. You can also use it to lay out straight lines between points that are within 30 inches of each other. Masons and cement finishers like to use a straightedge.

Spirit Level, Line Level and Level Sights

A spirit level is also called a carpenter's level or a mason's level. It's a straightedge made of wood, aluminum, magnesium, or plastic. It comes in lengths of 2, 4, or 8 feet. Use it to check whether a surface is level and plumb. To find horizontal and vertical planes, center a bubble suspended within a glass tube or vial parallel, or perpendicular, to the surface of the level.

A line level is a small metal and glass tube with a spirit bubble you can suspend from a line to show whether something is level or not.

Level sights have a peephole piece and a cross hair piece attached to a spirit level. When you set up the level horizontally on a solid base you can sight to a distant object to find out whether it's at the same level.

Miscellaneous Survey Equipment

Other necessary equipment for surveying is:

- Four-pound sledgehammer for driving stakes
- Claw hammer for building batterboards
- Pick hammer to dig rocky soil
- Machete to clear brush
- Bush ax to clear heavy brush
- Hand ax to sharpen stakes
- Hand saw to cut batterboards and posts
- Rock/concrete chisel to set lead-and-tack markers
- Concrete drill to set lead-and-tack markers
- Tacks, nails, and spikes for use as markers
- Flat shiners to make markers more visible

- Flagging or survey tape to make survey stakes more visible
- Crayon and spray paint to identify stakes
- Stake bag and tool pouch to carry equipment

Maintaining Horizontal and Vertical Control

On most construction projects, you must continuously maintain horizontal and vertical control over the work. Here are some suggestions on how to do this.

The key to maintaining horizontal control is the permanent survey monument. The monument should be unaffected by settlement and frost action. This marker is usually made of a concrete base with a bronze cap cast in the top. When a monument is set in a roadway or paved surface, a cast iron hand hole is often placed over the cap for protection.

Governmental monuments for control points on triangulation or traverse systems are usually made under First Order Survey accuracy. The monuments normally have a $3^1/_2$-inch-diameter red brass cap. The cap is engraved with letters and symbols describing the name, number, and type of monument. Some monuments are designed to be cast into concrete and others fit into cast iron sleeves that are cast in concrete.

When a construction project covers a large area and includes many buildings and other structures, it's a good idea to set up a coordinate system. To make this system establish a grid over the site, usually in 100-foot increments. If possible, the grid should be parallel to two property lines that are at right angles to each other, as it is back in Figure 6-5.

When a project occupies a relatively long and narrow site, the transit line is a good way to maintain horizontal control. This line is marked every hundred feet, or every station. The beginning of the line is called Station 0+00. Successive stations along the line are called Station 1+00, 2+00, and so forth. Typical points between stations are described as Station 1+32.21, Station 2+46.96, and so forth. Measure all key locations on the site a certain distance to the right or left and 90 degrees to the transit line.

This is the system you'll usually use to lay out the rough plumbing in large buildings. Sometimes, two transit lines set at right angles to each other are marked on the site. The lines are called "Line A" and "Line B." Then you can measure key points in the rough plumbing system as offsets from the respective transit lines.

Vertical control means establishing accurate elevations throughout a project, usually by referring to a government bench mark. This may be a U.S. Geodetic and Survey monument which shows its elevation as the distance above mean sea level.

You can set the project bench mark as Elevation 100.00. Then you should record an equation that states the equivalent USGS elevation. For example, Datum Elevation 100.00 = USGS Elevation 396.27.

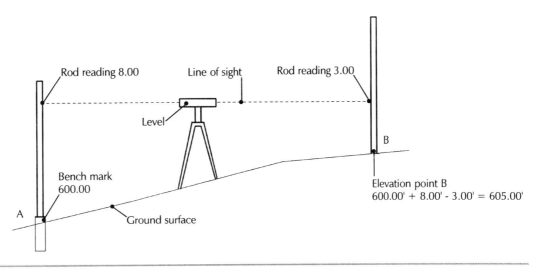

Figure 6-14 Using a level and leveling rod to set an elevation

To set elevations using the level and leveling rod, as shown in Figure 6-14:

1) Focus the telescope to read the figures on the rod.

2) Move the telescope to align the vertical cross hair to split the rod.

3) Center the bubble in the leveling vial between the two graduations.

4) Read the rod to the nearest one-hundredth of a foot.

5) Check the bubble to make sure it's still level.

6) Record the rod reading.

A popular method of establishing levels is to use the laser plane automatic level, as shown in Figure 6-15. One person, equipped with an automatic level, transmitter, receiver, and rod, can set grades for foundations and slabs.

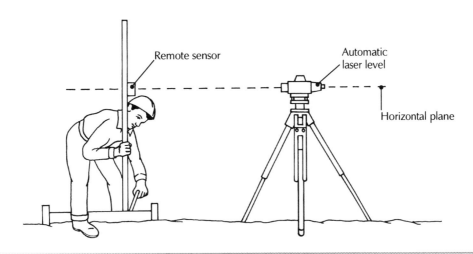

Figure 6-15 Using a laser plane automatic level

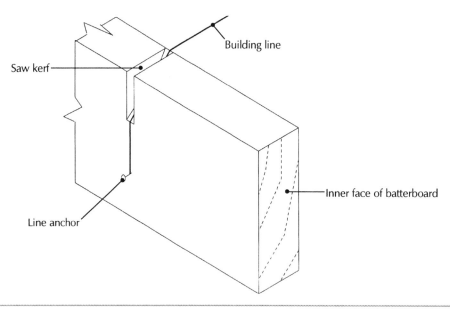

Figure 6-16 *Sawing a kerf in a batterboard*

Laying Out a Building

The first layout task is the foundation. You'll need the plot plan, floor plans, foundation layout, and elevation reference drawings. You'll also need three 2 × 4s, each from 4 to 5 feet long, for corner batter stakes, and two 4-foot 1 × 6s per corner board. Keep handy a supply of 6d or 8d common nails, mason line as long as the perimeter of the foundation, and 2 × 2 position stakes 12 inches long.

There are several good ways to lay out a building foundation. One simple method is to stake out a rectangle along the major outer dimensions of the structure. Then stake out the irregularities of the building shape as smaller rectangles.

Mark the building corners with corner hubs, right angle batterboards, and posts. Place batterboards 3 to 4 feet outside of a corner stake. Use 2 × 4 wood to make posts, and 2 × 4 or 2 × 6 lumber to make batterboards.

Set the top of the batterboard at the same elevation as the top of the foundation wall. Drive nails into the top of the batterboards and tie cords onto opposing nails to establish the building line. Where two strings cross, drive a 2 × 2 hardwood hub into the ground and hammer a nail into the top of the hub. This marks the exact location of the true corner. Another way to attach the cord to the batterboard is shown in Figure 6-16.

An alternate procedure for staking out a building is shown in Figure 6-17. The procedure is:

1) Locate or set property corner stakes. New corner stakes should be set by a licensed surveyor.

2) Run a line across the front of the property and along the two sides. Do all measuring from these two lines.

3) Drive stakes at the approximate location of building corners.

4) Measure from the front property stake alongside the property stake a distance equal to front setback, then drive a stake at this point.

5) Measure from this stake along the side property line a distance equal to the depth of the building from the front to the back, and drive another stake.

6) Repeat this operation along the other side property line.

7) Measure in from the front and rear setback stakes on the side property line a distance equal to the side yard, and drive stakes at the two points. This establishes two building corners.

8) Repeat this operation along the other side property line. This establishes the other two building corners.

9) Remove all building lines.

10) Install batterboards around each corner stake. Use three stakes and two horizontal boards set at right angles. Set the top of the batterboards at the same level as the top of the foundation, if possible. Set the boards above or below the top of foundation, if it's more convenient. Set the batterboards far enough from the building corners so they don't interfere with the excavation of the foundations.

11) Mark the exact corner locations on the batterboards.

Check the accuracy of a batterboard layout by measuring triangular distances, as shown in Figure 6-18.

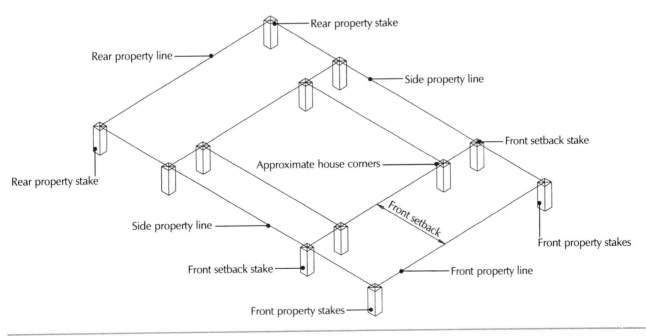

Figure 6-17 *Staking out a building*

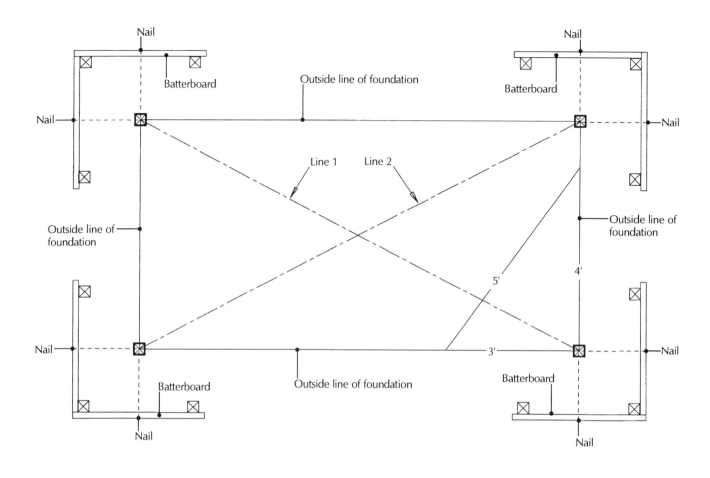

Figure 6-18 *Checking diagonals and angles for squareness*

Laying Out Interior Walls

Laying out of interior walls on the first floor slab or subfloor is the next important task. Carefully check the plans to decide whether the dimensions shown on the plans are to the center or face of the stud. If the dimension line is to the face of the stud, make sure which face. The usual practice is for a dimension to run from the outside face of exterior walls to the center of each interior wall. All intermediate dimensions should add up to the overall exterior dimension. When the dimensions are given in mixed fractions, convert all fractions to a common denominator. For example, you'd change 8'6^1/$_2$" + 3'4^3/$_4$" + 2'^{5}/$_8$" to 8'6^4/$_8$" + 3'4^6/$_8$" + 2'^{5}/$_8$". Adding and subtracting a string of dimensions is easier when they have similar fractions.

Sometimes, as in bathtub enclosures, the plans show the dimension to the outside surface of adjacent studs to assure the required clearance. For 2 × 4 stud walls, assume their depth to be 3^1/$_2$ inches. Where 2 × 6 or 2 × 8 studs are used in exterior walls or interior walls containing plumbing, deduct ^{1}/$_2$ inch from each stud dimension to find its actual depth.

Construction Tolerances

Dimensional tolerances in construction are very important. You can calculate a distance mathematically much closer than you can measure it physically in the field. In addition, each building trade has its own permitted tolerance. For example, steel construction, masonry construction, rough carpentry, and grading work have different allowable tolerances. Subcontractors may demand back charges when errors in field measurements exceed their standard tolerance.

Surveyors generally have three levels of accuracy. The first order is for geodetic governmental surveys. The second order covers land subdivision surveys. Second order surveys allow a tolerance of one part in 10,000, or 1:10,000. For example, a tract boundary line 1000 feet in length must close within 0.1 foot. A distance of 100 feet must be accurate to 0.01 foot. The third order is too approximate to be used in construction.

Structural steel fabrication and erection require a relatively close accuracy for placement of bolts. A bolt hole diameter is only $1/16$ inch larger than the bolt's diameter. The allowable tolerance between any two anchor bolts in a group of bolts is $1/8$ inch. The tolerance between the center of adjacent groups of anchor bolts is $1/4$ inch. The tolerance in a 100-foot column line is $1/4$ inch and the tolerance between the center of an anchor bolt group from the column line is $1/4$ inch. Figure 5-17 in the previous chapter shows the allowable error in setting anchor bolts for steel construction.

Some masons consider 1 inch as "close enough" for the foundation of a masonry wall. Manufactured concrete unit masonry (concrete block) allows a tolerance of $3/8$ inch in the thickness of the block. So, if a building line is the outside surface of a block wall, the interior surface may vary by $3/8$ inch. This would affect all dimensions taken from the interior face of a concrete block wall.

Grading work has its own standards. Rough grading is measured to $1/10$ of 1 foot accuracy while elevations of pavements are set at $1/100$ of a foot accuracy.

Concrete slabs on grade are considered to be a straight plane if there's less than $1/8$-inch variation in a 10-foot distance. In addition, set three consecutive points on the same slope so you can see any variation from a straight grade.

Reinforced concrete beams and elevated slabs are usually constructed with an upward camber to allow for future deflection from dead and live load. Therefore, concrete structural members may have an upward curve when there is no live load. You should consider the camber when measuring from the top surface of a beam or slab.

The field surveyor should allow for these tolerances when setting up lines and elevations. Steel work is fabricated to $1/16$-inch tolerance. That's less tolerant than masonry work. When steel beams are supported by masonry walls, you should

Fraction of an inch	Decimal of an inch	Decimal of a foot	Fraction of an inch	Decimal of an inch	Decimal of a foot	Fraction of an inch	Decimal of an inch	Decimal of a foot
1/16	.0625	.0052	3/8	.3750	.0313	11/16	.6875	.0573
1/8	.1250	.0104	7/16	.4375	.0365	3/4	.7500	.0625
3/16	.1875	.0156	1/2	.5000	.0417	13/16	.8125	.0677
1/4	.2500	.0208	9/16	.5625	.0469	7/8	.8725	.0729
5/16	.3125	.0260	5/8	.6250	.0521	15/16	.9375	.0781

Figure 6-19 *Conversion of fractions of an inch to decimals of an inch and a foot*

provide a means for field adjustment. Provide flexibility by using slotted holes or field welding. If the plans call for embedded anchorage in masonry walls without allowing for adjustment, the contractor should contact the designer of the building about a potential problem.

Units of Measurement

To lay out construction you'll need to convert several types of linear measurements. Architects use feet and inches to the nearest $1/16$ inch. Civil engineers use feet and decimal parts of a foot to the nearest $1/100$ of a foot. Road and sewer designers use stations that represent 100 feet subdivided into $1/100$ of a foot. Figure 6-19 shows conversions of fractions of an inch or foot to decimals of an inch or foot. You'll have to convert to the decimal system when you perform trigonometric calculations.

Different trades describe slopes in different ways. Carpenters use inches and fractions of an inch per foot (i.e., $3^1/2$ inches per foot). Plumbers use fractions of an inch per foot (i.e., $1/4$ inch per foot). Sewer contractors and civil engineers use decimal parts of a foot per foot or percent to indicate slope (s = .002 or .2%). Steel detailers use a method involving feet and inches to the nearest $1/32$ of an inch. Slopes are described by rise, run and slope using Smoley's Handbook rather than conventional trigonometry.

You may find angular dimensions in degrees, minutes and seconds, bearings, or azimuth. A line may be described as 15° 30'15" right of the preceding line. This same line may be described to be S 32° 23'10" W. Or it may also be described as having an azimuth of 212° 23'10".

Some angular directions are given in degrees and decimal parts of a degree. In steel framing, they describe the slopes of lines as the rise in a 1-foot run. Rise is measured in inches to the nearest $^1/_{32}$ of an inch.

The field surveyor must be able to convert the information indicated on plans of various disciplines into a common language. Because of the need for trigonometric computation, the common language for distances is usually feet and decimal parts of a foot. The conventional unit for angular measurements is degrees, minutes, and seconds.

In summary, the units of measurement used in construction are:

▌ *Angle:* This is the angular difference between adjacent lines given in degrees, minutes, and seconds. There are 360 degrees in a circle, 60 minutes per degree, and 60 seconds per minute. Degrees are shown on plans as ° , minutes as ' and seconds as ". For example, 45 degrees, 30 minutes and 15 seconds is given as 45° 30'15".

▌ *Azimuth:* Azimuth is the direction of a line measured in a clockwise direction from north. This system is used by surveyors and civil engineers in finding the closures of the boundaries of a property.

▌ *Bearings:* The direction of a line relative to north or south is called a bearing. A line 45 degrees west of north is given as N 45° W. All boundary lines of a parcel described by metes and bound are given in bearings. Figure 6-20 shows the use of bearings and distances to describe a boundary. Figure 6-21 shows how a bearing is measured from a north-south line.

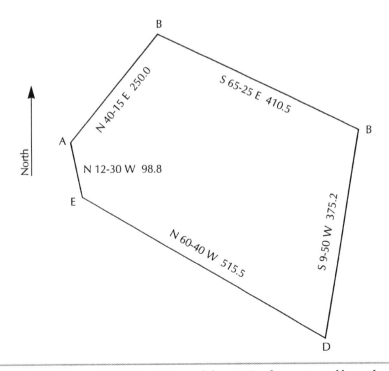

Figure 6-20 *Typical parcel description by metes and bounds*

- *Elevations:* Surveyors usually set elevations to hard durable surfaces such as concrete pavements in $1/100$ of a foot. Elevations of rough grading are normally stated in $1/10$ of a foot.

- *Feet and inches:* Buildings and structures are dimensioned by architects and structural engineers in feet and inches, to the nearest $1/16$ inch. Most manufacturers of building components such as doors, windows, and masonry units follow this system.

- *Feet and decimals:* Surveyors and civil engineers normally use feet and decimal parts of a foot, to the nearest $1/100$ of a foot. This simplifies the use of trigonometric calculations.

- *North:* True north and magnetic north are rarely used in any description of land subdivision. North in most legal descriptions is based on a previously-recorded document. Plot plans often show an arbitrary north arrow to make it easier to call out building sides as north, south, east, and west. This is sometimes called plant north.

- *Slopes and grades:* The direction of a line relative to a horizontal plane is given as a slope or grade. Units may be inches per foot, as used by carpenters and plumbers; decimal parts of a foot per foot, and percent of a foot per foot, as used by civil engineers and sewer contractors. Converting slopes from fractions of an inch to the decimal system is shown in Figure 6-22.

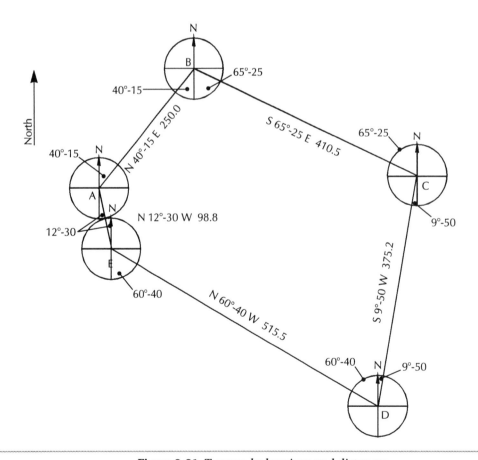

Figure 6-21 *Traverse by bearings and distances*

Inch rise per foot	Slope	Percent of slope	Inch rise per foot	Slope	Percent of slope
$^{1}/_{16}$	.0052	0.52	$^{9}/_{16}$	.0469	4.69
$^{1}/_{8}$	.0104	1.04	$^{5}/_{8}$	.0521	5.21
$^{3}/_{16}$	.0156	1.56	$^{11}/_{16}$	.0573	5.73
$^{1}/_{4}$	.0208	2.08	$^{3}/_{4}$	.0625	6.25
$^{5}/_{16}$	.0260	2.60	$^{13}/_{16}$	.0677	6.77
$^{3}/_{8}$	.0313	3.13	$^{7}/_{8}$	.0729	7.29
$^{7}/_{16}$	.0365	3.65	$^{15}/_{16}$	.0781	7.81
$^{1}/_{2}$	.0417	4.17	1	.0833	8.33

Figure 6-22 Conversion of slopes

Dimensions

To begin the layout of a building, you need the construction plans. Building plans usually consist of a plot plan, and architectural and engineering drawings. On larger, complex jobs, you'll have mechanical, electrical and plumbing plans, but you probably won't need them to lay out the buildings.

The plot plan shows the location of the building as it relates to the property lines. These plans are usually based on a surveyor's plat. Dimensions are normally given in feet and decimals of a foot, drawn to an engineer's scale (1 inch = 40 feet, for instance).

Architectural plans describe the buildings in detail and are drawn and dimensioned to an architectural scale. The scale is expressed in inches or fractions of an inch equal 1 foot (for example, $^{1}/_{4}$ inch = 1 foot). Dimensions are given in feet, inches and fractions of an inch (as in 12'5$^{3}/_{4}$").

Engineering plans describe the structural system of the building, such as reinforced concrete or structural steel. Details of reinforced concrete structures are drawn and dimensioned according to the standards of the American Concrete Institute (ACI), which are similar to architectural standards. Structural steel drawings are usually drawn following the rules of the American Institute of Steel Construction (AISC). Dimensions are also shown the same as on the architectural plans.

Certain governmental construction projects require dimensions in the metric system, plus the English system. Builders of private residential and commercial structures don't have to contend with this — yet. There's a possibility that by the year 2000 all building codes will be primarily in the metric system. The latest 1997 building codes are in both English and metric systems. Metric scales are expressed in ratios of the size of a drawn object (for example 1:50, where 1 meter on the drawing

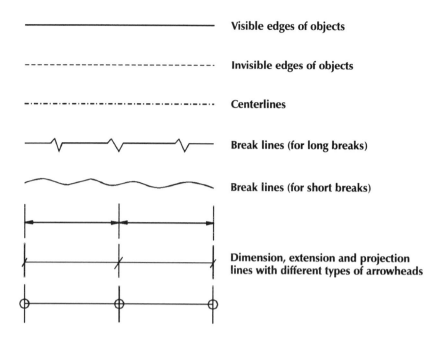

Figure 6-23 *Conventional types of lines on drawings*

represents 50 meters in the building). Metric dimensions are given in meters and millimeters (as 10.125).

For this book, we'll deal only with the architectural and engineer's scale. The layout person must merge the decimal system of the plot plan to the foot-and-inch system shown on the architectural plans.

Drafting Symbols

Lines and dimensions shown on the construction plans are the means of communication between the designer and the builder. That's why it's so important that the plans be clear and accurate enough so the layout person can understand the intentions of the draftsperson. In addition, the person doing the layout should check all the dimensions. If they don't add up, call the superintendent, architect or engineer.

There are many types of lines shown on plans. Some represent the visible edges of objects, invisible edges of objects, dimensions, extensions and projection lines, break lines, centerlines and radius lines. See Figure 6-23 for the common types of lines on drawings.

In addition, there are some differences between the way architectural and engineering plans are drawn. There's more variety in the style in architectural plans. Arrows on dimension lines may be a slash, dot or small circle, while engineering drawings generally use a small arrowhead.

Building layout lines usually represent the outside face on the building frame. Be careful to distinguish dimensions to the outside surface of a foundation wall from measurements to the exterior face of siding or plaster. Architectural plans sometimes show dimensions to the outside surface of exterior walls when the dimension is supposed to go to the face of the stud, not the exterior finish. The edge of a stud is usually the same as the edge of a foundation wall. See Figure 6-24 for typical dimensions on a townhouse plan.

Checklist for Construction Survey

Before beginning the layout, check that all the following items are properly represented on the plans. Also make sure that you know all the building and zoning code requirements. If there are any errors in the plans, now's the time to catch them.

Building Setbacks

This is the minimum required distance between the front property line and the front face of the building. Be careful to check the setback required by the tract restrictions and the local zoning code. They may be different. Subdividers sometimes require larger setbacks to enhance the value of the property.

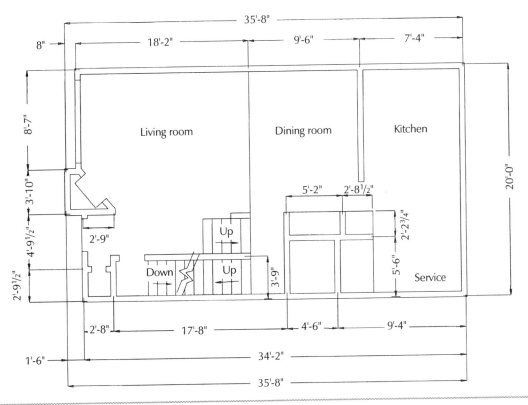

Figure 6-24 *Typical dimensions of a townhouse*

Side Yards

This is the minimum distance required between the side property line and the outside surface of the exterior wall. Side yard distance usually increases with the height of the building. Roof overhangs and eaves may project a limited amount into the side yard. This distance may vary between the main building and auxiliary buildings, such as garages.

Rear Yards

This is the minimum required distance between the rear property line and the rear face of the building. This distance may also vary between the main building and auxiliary buildings, such as garages.

Topographic Survey

On any construction work on a building site that's not level, you should have a plan showing the existing and proposed contours of the land. Usually, a licensed surveyor does a topographic survey and prepares a topographic plan showing the property lines, existing contours and physical features on the site and adjoining public works. The public works include the roadway, parkway, sidewalk, utility poles and other publicly-owned features. The building designer, architect or engineer uses this plan to design the grading, drainage, sewerage and buildings.

Orientation of Buildings

The orientation is the location of the building in respect to the property lines. In multiple building projects, such as housing or condominium developments, a single dwelling plan may be used for various locations. Each may have a different orientation. Be careful when you see on the floor plan a note reading "opposite-hand," "reversed," or "mirror image."

Exterior Dimensions of Buildings

Building sizes noted on a plot plan usually include the exterior finish. This may be greater than the architectural plans that are dimensioned to the exterior surface of the wall studs. Watch for this. A 1-inch difference in a side yard may be the cause of dispute with the building inspector.

Bench Marks and Elevation Reference

This is the reference for establishing vertical control of the building. The plot plan should show at least one elevation point that's marked in the field and visible from the building site. You'll base the elevation of the first floor on this point.

Elevation of Property Corners

These elevations should be shown on the plot plan, grading plan or topographic plan.

Curb and Gutter Elevations

These elevations should also be shown on the plot plan, grading plan or topographic plan. They're important in the design of the site drainage.

Finished Elevations of Paved Areas

The plot plan or grading plan should show these elevations. They're important in the maintenance of proper drainage. The layout of the edge form boards is set by these elevations.

Finish Floor Elevations

These elevations are important in establishing upper floor elevations, plumbing outlets and adjacent finish grades.

Centerline and Widths of Roads

The adjacent road centerline and width of the roadway should be shown on the plot plan. Check with the local street department whether there are any plans for future widening of the roadway. If there are, it may be the property owner's responsibility to build the future curb and sidewalk when the building is constructed.

Coordinate System

A grid system is useful on large projects to maintain horizontal control of all construction. You should lay out the grid coordinate system and mark it with monuments before you start the construction layout.

Utilities

The plot plan should show the location and description of all utilities. All private construction is dependent in some ways on public utilities. There may be an easement along the rear or side yards for power poles, underground pipes or electric conduits. The surface of the driveway must join with the sidewalk. Storm drains must join with the street gutter or main storm drains. Waste lines must connect to public trunk sewers.

Chapter 7

Workmanship, Plans, and Coordination Between the Trades

Basic wood engineering requires a knowledge of good workmanship in wood framing. You need to know industry standards for the quality and character of the finished work. The wood framer, like other contractors, is expected and required to perform his work skillfully and well. In fact, most construction contracts require that all subcontractors (including the framer) warranty their work for one year, replacing defective items and correcting any problems caused by poor workmanship.

Some of the basic rules for good workmanship are as follows:

❚ Install all work under the direction of experienced construction supervisors.

❚ Work safely at all times to protect life and property.

❚ Erect all framing true and plumb.

❚ Erect temporary bracing as necessary to support all loads placed on a structure and leave the bracing in place until a permanent structure is in place.

Dimensioning

Good workmanship starts with accurate dimensioning. Dimensions must be accurate and complete to properly lay out the walls. Figure 6-17, in the previous chapter, shows how to stake out a simple house on a regular lot. It's customary to measure from the outside surface of the exterior studs to the center of interior studs. Make sure that strings of dimensions add up to the overall dimensions. Locate walls that are at an angle to each other by the number of degrees in the angle. Figure 7-1 is a typical building foundation layout.

Take special precautions to locate anchor bolts and hold-downs, since they're embedded in concrete, where there's very little tolerance for error.

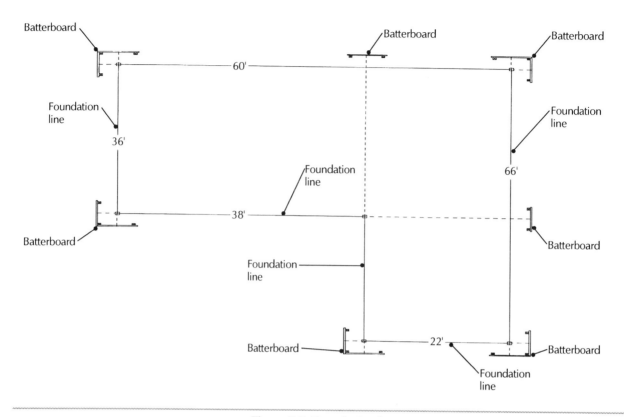

Figure 7-1 *Foundation layout*

Construction Documents

Good workmanship requires that you comply with the construction documents, including addenda, revisions, and change orders. The basic documents include:

▌ *Blue-line drawings* such as site, architectural, structural (framing), mechanical, plumbing, and electrical plans. Figure 7-2 shows a portion of a typical architectural floor plan for a condo unit. Dimension and extension lines are shown, but dimensions are omitted for clarity.

▌ *Specifications* which describe the scope of work, quality of materials, and quality of workmanship required for each trade.

▌ *Construction agreement or contract* which establishes the responsibilities of the contractor. Subcontracts establish the responsibilities of the subcontractors.

▌ *Shop drawings* of prefabricated wood components, which are submitted by the fabricator for approval by the owner, builder, engineer, and/or architect.

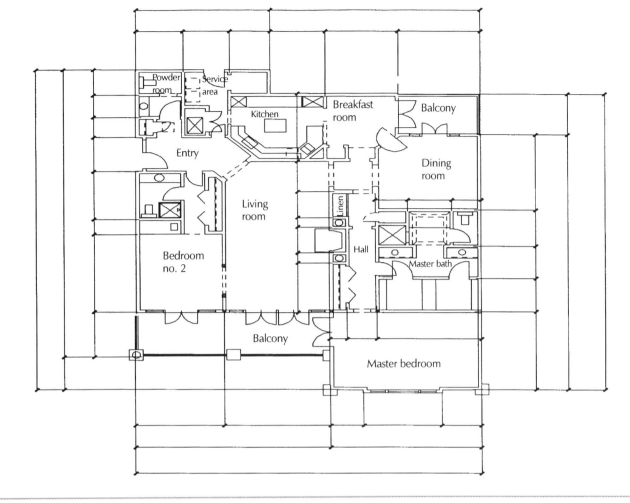

Figure 7-2 *Typical floor plan*

Whenever there's a conflict between these documents, priority is usually given in the following order:

1) Construction agreement

2) Specifications

3) Approved shop drawings

4) Blue-line drawings

The subcontractor and general contractor sign the construction subcontract. On rare occasions the owner and lender reserve the right to review and approve subcontractors and will also sign a subcontract. By signing this document, the subcontractor usually agrees to:

▮ Comply with the construction documents, building department and inspectors.

▮ Comply with the edition of the building code that was current when the permit was issued.

- Guarantee the work for a specified period of time (usually one year).

- Abide by all safety regulations.

Some of the major requirements usually specified in a subcontractor's agreement are:

- Provide a continuous quality-control inspection.

- Clean up regularly and store debris in containers at locations specified by the owner.

- Follow the project schedule to avoid delays.

- Assign the contract only with the approval of the general contractor and the owner.

- Coordinate job activities with other trades.

- Correct any and all defective work.

Structural Framing Plans

Structural framing plans are usually drawn over "brown lines" or "transparencies" of the architectural roof and floor plans. Figure 7-3 shows a typical framing plan for the condo unit floor plan shown in Figure 7-2. Framing members are drawn in heavy broad lines so they're easily seen. Although information shown on the structural plans overrides the architectural plans, the two sets of plans should be in agreement.

Framing plans show the location, size, spacing, and direction of all repetitive members such as rafters and joists. Single structural members such as beams, headers, lintels, girders, and posts should also be shown. Finally, the plans should indicate the locations of all mechanical devices and other elements required to resist wind or earthquake. These include tie straps, hold-downs, drag struts, and diaphragm chords. Wall dimensions don't need to be shown on the framing plans since they're drawn on the architectural plans.

The designer can save time writing specifications by referring to recognized authorities regarding wood framing. Here are some suggested notes:

- All wood framing shall be in conformance with the *Uniform Building Code,* 1997 or latest edition.

- Plywood shall conform with applicable standards listed in Chapter 60 of *Construction and Industrial Plywood-Product Standard PS-1-94* of the U.S. Department of Commerce, National Bureau of Standards.

- All lumber and plywood shall be identified by the grade mark or certificate of inspection issued by an approved agency.

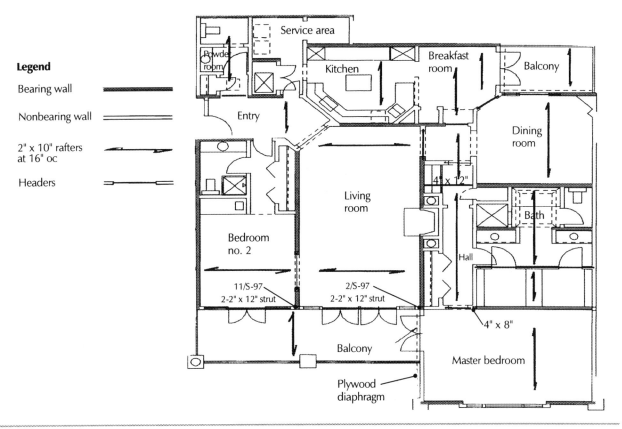

Figure 7-3 *Typical framing plan*

The framing plan or specifications should specify the grade and species of all wood members, as well as the manufacturer, model number and type of framing hardware. Schedules for nailing horizontal diaphragms, shear walls, tie straps, hold-downs, and other types of connections should be shown on the framing plans. See Figure 7-4.

Details showing connections of all the critical members of the framing system are a very important part of structural drawings. Critical members include girders, beams, posts, trusses, shear walls, and diaphragms. Figures 7-5 and 7-6 show two typical details of industrial roof connections.

Custom Plans

Plans are generally classified as "custom plans" or "builder's plans." Custom plans are usually prepared by a design team consisting of an architect, and mechanical, plumbing, electrical, civil, and structural engineers. Custom plans are more detailed and come with complete specifications. Every important architectural and structural connection is detailed. Very little is left to the subcontractor to design. Specifications name manufacturers and quality of materials.

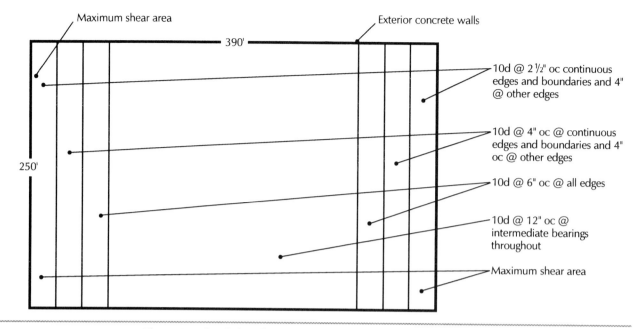

Figure 7-4 *Roof plan showing nailing pattern of plywood panels*

Builder's Plans

Builder's plans are much simpler than custom plans. They usually show only basic room and building layout, minimum specifications, and just enough information to get a building permit. Architectural and engineering fees are lower, since there's less time spent on the design. Contractors have more freedom in bidding the job because they have more flexibility in selecting materials and methods, since they may not be specified. Sometimes, the only test is whether "it meets the code." Construction usually costs less and contractors have more latitude to make changes easily as the work progresses.

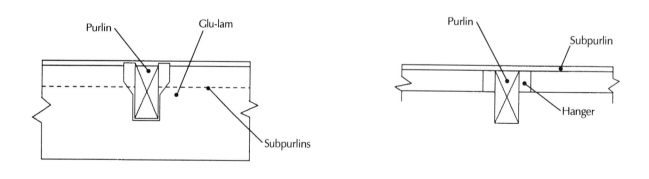

Figure 7-5 *Subpurlin to purlin connections*

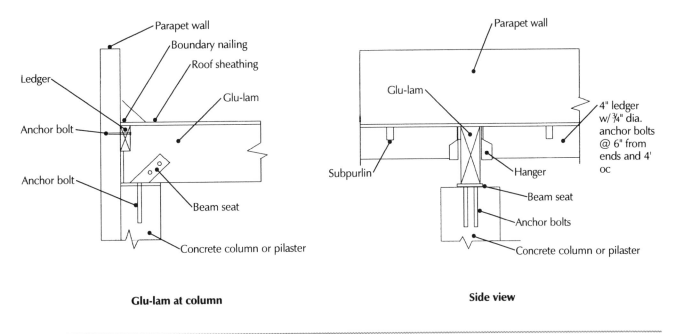

Glu-lam at column   **Side view**

Figure 7-6 Purlin to glu-lam connection

Projects built with builder's plans seldom require a full-time superintendent. This saves money but is more likely to result in problems. Delays often occur because of the lack of coordination of different trades. Records of changes and revisions are rarely kept and as-built drawings are usually nonexistent. Change orders are often sealed with a handshake. As a result, this type of project often leads to an unhappy owner, builder, or subcontractor filing a lawsuit at the local courthouse.

Framing Specifications

Wood framing specifications are as important as the architectural and structural drawings. The designer is responsible for the contents of the carpentry section of the specifications. This section describes the quality of materials and workmanship required in wood framing. It should include the following divisions:

Workmanship

A. *Scope of Work*

Work to be performed under this Section of the specifications consists of all rough carpentry work as indicated on the applicable drawings, and as required for performance of other divisions of this specification. Builder shall furnish all materials, labor, equipment and tools necessary to perform this work.

B. Materials

1. Rough Lumber

Lumber shall be grades in accordance with the West Coast Lumberman's Association Standard Grading and Dressing Rules, No. 15.

All framing lumber shall be Douglas Fir, Construction Grade or Better, and each piece shall be individually marked.

All blocking, bracing and ties shall be of the same grade as used in the framing.

C. Workmanship

1. Rough Carpentry

All framing shall conform with the requirements of the governing Building Department, Uniform Building Code and local governing authorities.

Framing lumber and other rough work shall be properly cut, accurately set to required lines, plumb and level, and rigidly secured in place.

Studs shall run full height of all partitions, from sill to plate, except where studs shall run to top plates set under 4 × 12 headers.

Studs shall be drilled for the passage of pipes and conduit.

Nailing shall be done in a manner that will develop the safe working stress of all members.

Where bolts are used, holes shall be drilled the nominal diameter of the bolt, and washers shall be installed where head or nut bears on wood.

Follow framing details shown on structural drawings.

2. Plywood Shear Walls

Location of all shear walls shall be as indicated on the unit plans with a symbol (A, B, C, D, etc.). The letter indicates the type of plywood, nailing, tie-downs and hold-down force shown on the Shear Wall Schedule.

Framer shall install all hold-downs in concrete slab and floor framing.

Use solid sawn 4 × 4 or 4 × 6 posts at each hold-down as noted on the Shear Wall Schedule.

Hold-downs and tie straps shall have the capacity equal to or more than the hold-down force noted for the seismic analysis for each shear wall.

Tie straps shall be located at every door and window opening in shear walls.

All hardware used in shear walls shall be approved by governing authorities or the building department.

All studs to which two panel edges are nailed shall be a minimum nominal 3 inches wide.

Recommended Framing Procedure

The framing contractor is responsible for installing and bracing the building frame. Here are some installation guidelines:

▌ Always refer to the structural plans to determine the proper location of wood members.

▌ Coordinate the architectural plans for dimensions and the framing plans for the sizes and locations of the framing members.

▌ Don't install a damaged or inferior framing member.

▌ Never alter any prefabricated structural member *without* the written approval of the designer.

▌ Allow the fabricator to build members only after shop drawings have been approved by the designer and the general contractor.

▌ Never stack heavy construction material on the frame until it's properly braced.

▌ Store construction material in small stacks over bearing walls and never at midspan of joists or rafters.

▌ Avoid prolonged concentrated loads by distributing materials over roof or floor as soon as possible.

▌ Frame all members with square, tight joints that are properly connected with nails or bolts.

▌ Install all horizontal members subject to bending with the crown edge up.

▌ Don't splice horizontal members subject to bending between supports.

▌ Install all members so they rest on supports without being shimmed.

▌ Make sure board faces are parallel and flush with finish work such as drywall and sheathing.

▌ Whenever possible, avoid cutting or boring pressure-treated timbers. Coat all cuts and borings with an approved sealant.

▌ Consider wood shrinkage before you make a cut or bore holes.

Inspections

Most wood forming and framing jobs are inspected by city inspectors. You can't pour concrete until the trenching, forms, and reinforcing steel have been inspected. Don't pour a concrete floor slab until the underslab electrical, plumbing, and gas piping have been inspected and approved. And don't install wall coverings until the building frame, the rough plumbing, electrical, gas and HVAC are all inspected and approved.

In some metropolitan areas, the quality of inspection has diminished because of cutbacks in building departments. A recent study in one large city indicated that each building inspector was required to inspect 12 projects a day, averaging 20 minutes per site. You should realize that you aren't relieved of liability simply because the city has approved plans and construction. The city's "Approval Stamp" usually carries a waiver indicating that the approval is limited and doesn't apply to nonconformance to the building code. You're still ultimately responsible for the quality of your workmanship.

Coordination Between Trades

The quality of a wood frame affects practically all other subcontractors' work. When wood framing is installed incorrectly, other trades will suffer. Here are a few examples of the results of poor formwork and framing:

▌ Concrete slab out-of-square or out-of-level.

▌ Roof planes incorrectly sloped or rafters installed off-of-center.

▌ Roof parapets built too low for proper flashing and roofing installation.

▌ Missing roof crickets causing rainwater to pond on flat roofs.

▌ Lightweight concrete floor decking installed too thin because floors weren't framed level.

▌ Floor tiles cracked because the plywood subfloor wasn't nailed securely.

▌ Balconies drain back to doors due to improper slope of floor joists.

▌ Plastered walls uneven due to warped or bowed studs.

▌ Doors and windows not fitting in rough openings that were framed incorrectly.

▌ Elevator shaft rails not vertical because shaft wasn't framed true and plumb.

▌ Finished flooring wavy due to uneven floor joists and subfloor.

▌ Buoyant and squeaky finish flooring because subfloor wasn't nailed securely.

▌ Interior wood paneling coming loose because studs were spaced off-of-center.

▌ Sheet metal flashing leaking because of missing or misplaced wood backing and nailers.

▌ Loose cabinets because wood backing was missing.

▌ Plumbing difficult to install due to the lack of pipe space in plumbing walls.

- Shower and tub stalls don't fit because wood framing was installed incorrectly.

- Air conditioning not installed correctly because the wood platform for roof-mounted equipment was built in the wrong place or with insufficient height.

- Electrical fixtures not fitting because framing for recessed fixtures was missing.

Summary

The wood framer is the most important person in the construction team. The manual skill of the framer affects the quality of work of all the other disciplines on most projects. The form maker makes certain that the concrete work is accurately placed. The trades that follow the framer, such as the plumbers, tin-knockers, roofers and plasterers, depend on the workmanship and skill of the framer. Even the building designers, the architect and engineer, rely on the framer to provide the required strength and appearance to the building.

The final chapter in this book will introduce you to a simple computer program called *Wood Beam Sizing*. You can quickly learn to use it to calculate wood beam sizes for all of your projects. And check out the appendix, which covers building codes and standards, names and addresses of various building and lumber associations, and a glossary of wood and engineering terms.

Chapter 8

Using the Wood Beam Sizing Program

Wood Beam Sizing™ is a wood beam and post sizing program. It requires an IBM-compatible personal computer, and:

▌ Microsoft *Windows* version 3.1 or later

▌ 80386 or 80486 or Pentium microprocessor

▌ Color VGA screen

▌ Mouse

▌ Minimum of 2 Mb of free memory

Installing Wood Beam Sizing

Wood Beam Sizing is supplied on a 3¹/₂-inch disk in the back of the book. The disk contains a complete copy of the basic program. The program installs the same as any *Windows* program. Since the files are compressed, you must use the setup program to install. Copying directly to your hard disk won't work.

All Wood Beam Sizing files must be in one directory, except for the *Windows* system files, which are added to your Windows directory. We suggest directory C:\WOOD, although you may select any other directory you wish when you install the program.

To Install

- Start *Windows.*

- Insert the program disk in the disk drive.

- **For Windows 3.1,** at the Program Manager, select Run from the File menu.

- **For Windows 95,** press the Start button and select Run.

- Type A:SETUP or B:SETUP as appropriate and press Enter.

- Follow the setup instructions.

The Craftsman Book Company edition of Wood Beam Sizing (Version 1.31) does not contain some of the advanced features. At the Window menu, make sure *Basic* and *Start Up Notes* are checked. You can check *Advanced* to show what's available, but it's not fully functional.

References

The program is mainly based on the National Design Specification for Wood Construction (NDS). This national standard is published by the American Forest & Paper Association (formerly National Forest Products Association). Its latest edition is ANSI/NFoPA NDS-1991, which can be purchased at your local construction bookstore, or write:

American Forest & Paper Association
1111 Nineteenth St., NW, Seventh Floor
Washington DC 20036

You should use NDS in conjunction with your local building code: the Uniform Building Code (UBC), the International Building Code (IBC), the BOCA National Standard Building Code or the Standard Building Code (SBC).

Warning

Wood Beam Sizing is an educational computer program that's designed to handle basic mathematics. You make the design choices; the program will show you the consequences. You'll quickly get a feel for the general range of what's practical and what isn't.

Please note that no computer program can take the place of professional advice. Although this program has been prepared with reasonable care and uses standard engineering formulae, it may be very difficult to determine exactly what loads the beam may actually be carrying, especially for complicated roof lines or unusual floor framing.

There are many factors involved in the actual design of wood beams, especially larger beams, longer spans, connections, cantilever beams and concentrated loads. A computer program can't take the place of professional experience and judgment. If you have any questions about the results of this program, please consult your local building inspector, architect or engineer. Neither Craftsman nor NorthBridge Software Inc. intends to give professional advice and they accept no responsibility or liability in any manner for how this program may be actually used.

Learning to Use Wood Beam Sizing

The easiest way to learn Wood Beam Sizing is to follow through some examples, step by step. The program will show the results you get, but don't hesitate to experiment with any of the control window values. If by some remote chance you succeed in crashing the program, you'll be left back at the Windows Program Manager. No harm done. Just start the program again.

Basic Steps

These are the basic steps that we follow for sizing any beam:

▌ Select the wood species and grade.

▌ Select the span.

▌ Check the loading.

▌ Select the beam cross-sectional dimensions.

▌ Review the numbers.

We'll work through several examples following these basic steps.

Floor Joists

We're going to check out the span and stiffness of some 2 × 10s we have on hand. They're 16 feet 4 inches long. According to the grade stamp on the lumber, these are Hem-Fir No. 1 and 2 material.

First, we'll start Wood Beam Sizing. Whenever the Notes box opens and you don't want it, just close it to get it out of the way. Now press the *Strength* button at the top of the screen. Make sure the *Dimension Lumber* category at the lower right is checked, and choose *Repetitive Members* under *Adjustments*. We'll scroll through the choices and find Hem-Fir North No. 1 and 2. Click on it.

We checked the NDS and our local building code to find the allowable fiber stress in bending (Fb), and allowable shear stress (Fv) and the Modulus of Elasticity (E). The 1991 NDS, the 1997 UBC, the 1993 BOCA and the 1994 SBC give Hem-Fir No. 1 and 2 the following values:

Fb = 1000 psi
Fv = 75 psi
 E = 1,600,000 psi

After double-checking that *Dimension Lumber* and *Repetitive Members* are checked, hit the *OK* button. The current working strength values appear at the bottom of the screen, including the 15 percent increase for repetitive members (1000 × 1.15 = 1150 psi).

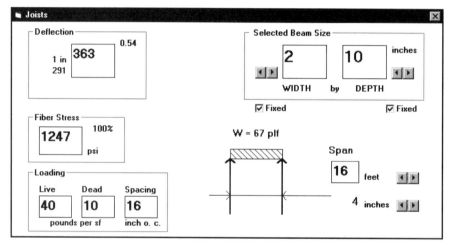

Actual size: 1.5 × 9.25 inches   Fb 1250   Fv 75   E 1.6   Hem-Fir-North No. 1 and 2

Now press the *Joists* button. Adjust the span to 16 feet 4 inches. Set the size to 2 × 10 and check Fixed under both the width and the depth. Click *Recalculate*.

For standard floor framing, this looks all right. The *Deflection* is 1 in 363 under live load (which is close to the usual 1 in 360 for floor framing) and the *Fiber Stress* is exactly right.

(Note: The Fb has been automatically increased by a 10 percent size factor for 10-inch deep lumber.)

But we decide we want a stiffer floor. Let's look at the current suggested 1 in 480 minimum floor deflection. We reduce the span down to 15 feet and try it. Let's try again at 14 feet 10 inches. Click the *Notes* button. It reads: "Looks pretty good!!!"

A Roof Beam

We decide we're going to have an open volume ceiling (a "cathedral" ceiling) on the second floor of our new house. We'll have a roof beam, running along the ridge, to support the roof.

First, click on the *Loading* button. Then press the *Weight* button to check out the live and dead loads for our problem. The default weights look all right.

Loading

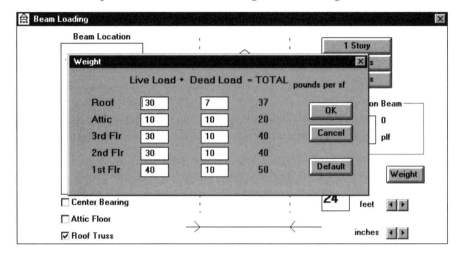

Now click on the *OK* button to get back to the *Beam Loading* window. Click on the *Roof Truss* box to turn it off and the *Center Bearing* box automatically goes on.

(Note: If you don't have either a roof truss or an attic floor, then the roof is unstable. You must have a center bearing wall or posts to support the roof.)

Our house is 24 feet wide, so we don't need to change the *Building Width*.

Finally, at the *Beam Location* selection box, we choose the location for our roof beam, right in the center, under the ridge. Check the second box on the Roof line under *Beam Location*.

Loading

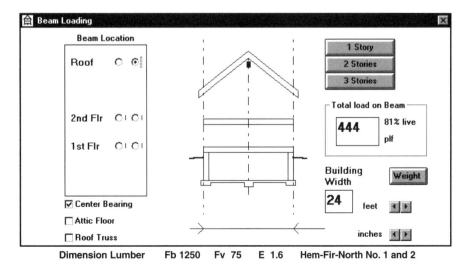

Now press the *Notes* button at the top to check out the choices.

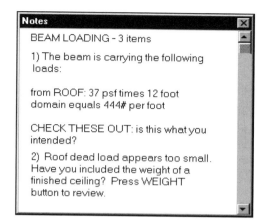

Note #1 looks right — we're supporting half of the roof. But Note #2 reminds us to press the *Weight* button again.

Press the *Weight* button. Change the roof dead load to 10 psf, to include the gypsum wallboard ceiling. Press OK and select the beam location again. Our *Total Load on Beam* now reads 480 pounds per linear foot (40 × 12 = 480).

We want to use our Hem-Fir 2 × 10s to make up the roof beam. (See Example 1 if you haven't set the Strength window correctly.) Press the *Beam* button. Then click on the *Fixed* box under the beam *Depth* and adjust it to 10 inches. Press *Recalculate*.

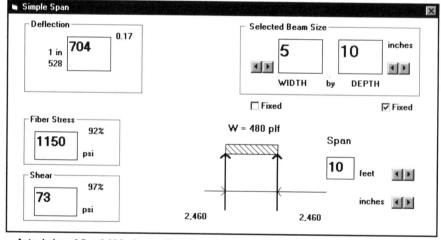

The beam size is 5 × 10, so we can use three 2 × 10s spanning 10 feet. The *Actual Size* appears briefly in the lower left corner of the wood strength line on the bottom of the screen: 4.5 inches wide by 9.25 inches deep.

The support posts at each end must handle a total load of 2,460 pounds each, which includes the estimated weight of the beam.

But the *Fiber Stress* is 92 percent of allowable. How much greater can we make the span without exceeding the fiber stress? Here's a hint: Not much.

Suppose we want to increase the span to 16 feet along the ridge beam between posts. How many 2 × 10s do we need? (The bottom of the screen shows 11.5 × 9.25, so it takes about eight.)

When you have the combination of span and size that you want, press *Notes* to review your design. Does it look right?

From the File menu, click *Print Results*. Be sure that the window you're printing from is the most recent active window, and the *Recalculate* button has been pressed.

Simple span beam, uniformly distributed load

Loading: w = 480 pounds per linear foot, total load on beam
        Live load: 75% of total load

Beam span: 16 feet 0 inches center line to center line

WOOD TYPE to be used is Hem-Fir-North No. 1 and 2

Working fiber stress has been adjusted for:
        Repetitive members - 15% increase
        Size factor - 1.10
        (for beam 10 inches deep)

MAXIMUM PERMITTED working stresses are therefore:
        Fiber stress in bending: Fb = 1,250 psi
        Horizontal shear: Fv = 75 psi
        Modulus of elasticity; E = 1.6 million psi

BEAM SELECTED based on the above information:

        12 inches nominal width
by
        10 inches nominal depth

Beam will deflect 1 in 439 under live load

Actual fiber stress under full load: 1,194 psi
Actual shear stress under full load: 48 psi

When you exit Wood Beam Sizing, your selection of wood strength and weights are saved for next time.

Full Size Beams

We're renovating an older house. The first floor is framed with full 5 × 5 Red Oak beams, 40 inches on center, that span 10 feet. How strong are they?

First, we'll go to the *Strength* window. Select *Beams and Stringers*, not *Dimensional Lumber*. Then select Red Oak No. 1 as the closest choice.

Strength

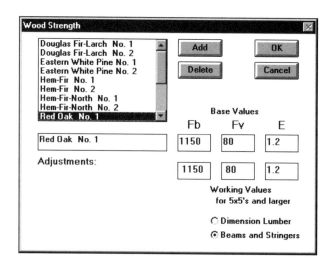

Press *OK*. These are full size timbers. We need to set the theoretical nominal or rough size. Click on the *Joists* button. We get a reminder that the Repetitive Member factor usually applies to joists, but in this case they're spaced too far apart. So click on *OK* to ignore the reminder.

Now set the *Width* and *Depth* both *Fixed*, then enter 5.49 for width and 5.49 for depth. (We're assuming a rough size of 5.49 by 5.49.) Use the mouse and the keyboard numbers to set the dimensions, not the scroll arrows on the screen. Set the *Span* to 10 feet and the *Spacing* to 40 inches. Now press *Recalculate*. You'll get a message that the beam is overstressed. Click on *OK*.

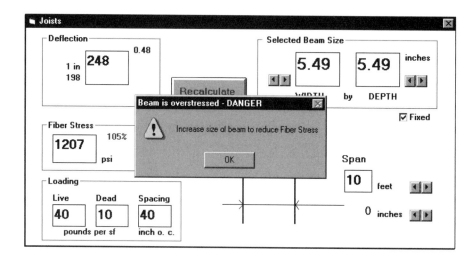

It looks like the existing floor loading at 40 pounds per square foot live load is strong enough for the first floor. The fiber stress is only slightly above the allowable strength of 1150 psi. But the deflection looks like a problem. The joists will sag almost 1/2 inch (0.48) under live load. The actual size line at the bottom of the screen shows the actual dimensions of 4.99 × 4.99 inches.

If we want the fiber stress exactly 1150, what would the live load have to be? Let's try a live load of 35 pounds per square foot. This gives us a fiber stress of 1086 psi. Not quite enough. Try a live load of 37.5 psf. That's it — the fiber stress is 1146 psi.

The second floor is framed the same way. If we use a standard live load of 30 psf for the second floor, is the deflection any closer to the 1 in 360 we would like for floor framing? If the live load on the second floor is 30 psf, the deflection calculates out to 1:331.

Another room on the first floor has an 8 foot 10 inch span. Is this OK at 40 psf? Yes — the deflection is 1:360.

Wood Posts

To calculate axial loading and beam end compression, click on the *Posts* button.

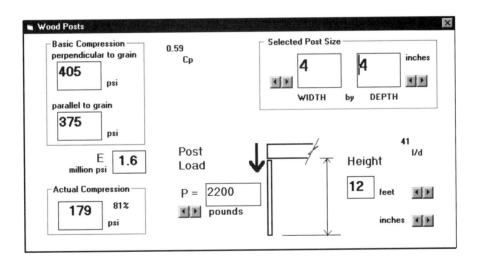

The axial load is set here at 2,200 pounds, with an unbraced post height of 12 feet. The Basic Compression values of 405 psi perpendicular and 375 psi parallel are for Hem-Fir No. 2. Also set the E value at 1.6 for Hem-Fir. You must manually enter the basic compression values for the wood you're using. If the beam is of a different material than the post, enter the beam species value for perpendicular compression (beam end crushing) and the post species for parallel compression (axial loading).

In this example, the calculation for a $3^1/_2 \times 3^1/_2$ post is:

2,200 pounds divided by 12.25 equals 179 psi, actual compression, or

$$2{,}200 \div 12.25 = 179 \text{ psi}$$

At a length-to-depth ratio, l/d, of 41 and C value of 0.59, the allowable maximum compressive stress is calculated to be:

$$0.59 \times 375 = 221 \text{ psi (which is less than 405, and governs)}$$

179 psi is 81 percent of 221 psi. It's OK.

Caution: No allowance for eccentricity is included.

Basic Wood Design

The basic steps in designing a wood beam are:

Beam Loading

▌ Determine all dead and live loads supported by the beam.

▌ Determine where they occur, concentrated or distributed. We can only consider distributed loading in this basic program.

Wood Strength

▌ Select the type of wood to be used.

▌ Make any adjustments needed for design conditions.

Configuration

▌ Select a single timber (Beams or Stringers) or a built-up beam (several pieces of Dimension Lumber).

Beam Sizing

▌ Select the size of the beam.

▌ Are bending stress, shear stress and deflection OK?

Bracing and Support

▌ Is the beam adequately braced?

■ Is there sufficient bearing at each end?

■ Are the loads carried successfully to the ground?

Added Safety Factor

■ How sure are you of your design assumptions?

■ How sure are you of the quality of construction?

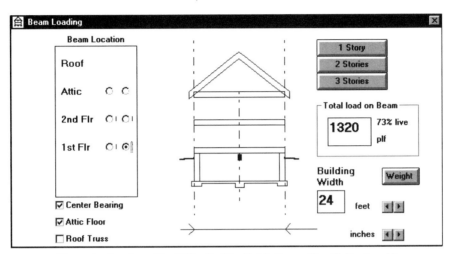

Dimension Lumber   Fb 1000   Fv 70   E 1.4 Spruce-Pine-Fir No. 1 and 2

The Windows

Beam Loading Window

A beam supports both live and dead loads from the structure above. (Note: This is the hardest step — to actually determine how much of the building is really supported by the beam.)

To help you visualize the loads, select:

■ Overall Building Width. This is not the beam span. It's the size of the structure at right angles to the beam span that determines the tributary area (domain) of the loads.

■ Number of Stories: How many stories does the beam support?

■ Roof framing - a standard Attic Floor or a Roof Truss. The roof truss bears only on the outside walls. An attic floor is usually supported by a center bearing wall.

■ Is there attic storage or the possibility of turning the attic into future rooms?

▌ Center Bearing or not? Will loads from the middle half of the building eventually be supported by a series of basement columns and girders? Or does the floor span from outside wall to outside wall? Are there floor trusses? As part of this step, press the *Weight* button to review the live loads and dead loads. These are usually specified by the building code.

Adjust the square foot weights for your actual construction and conditions. Wood Beam Sizing will remember your choices for the next time.

Wood Strength Window

The strength of a wood member depends on the species and its quality. The species and grade is usually stamped by the mill on each piece of lumber. You can find their allowable bending and shear strength as well as their modulus of elasticity in your building code or the NDS.

Strength

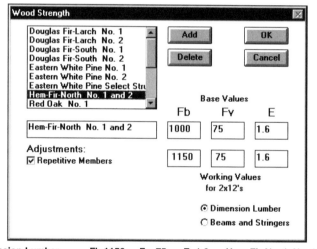

Dimension Lumber   Fb 1150   Fv 75   E 1.6   Hem-Fir-North No. 1 and 2

Wood Beam Sizing lists many common values already, but you need to be sure that these match the lumber you design for. You also need to decide here whether you'll be using Dimension Lumber or Beams and Stringers.

▌ Dimension Lumber: Standard dressed lumber, 2, 3 or 4 inches thick and up to 16 inches wide. Will you be building up the timber out of several pieces to make the required thickness? If you're using three or more pieces, select *Repetitive Members.*

▌ Beams and Stringers: Sawn timbers, 5 inches minimum thickness and up to 24 inches wide. Standard sizes are slightly different from dimension lumber.

Repetitive Member Factor

Repetitive members include joists, truss chords, rafters, planks, decking or similar members, which are in contact or spaced not more than 24 inches apart, and are joined by load distributing elements. The base bending values entered above will be increased by 15 percent if you check the *Repetitive Member* box. Each piece of wood has certain defects which reduce its overall strength. By using three or more members together, the same defects will not occur in the same place. So we get a 15 percent strength increase.

With joists and rafters, the subfloor or roof decking ties the members together so they can be considered repetitive members. When three or more 2 × 12s are spiked together to make a single beam, for instance, the same 15 percent strength bonus also applies.

(Note: The program automatically makes adjustment for the difference in actual size between Dimension Lumber and Beams and Stringers. You'll see the adjustment in the *Wood Strength* line at the bottom of the screen.)

Other CDs and videos are available from the International Conference of Building Officials (ICBO):

1997 Uniform Building Code on CD-ROM Version 1.3

2000 International Residential Code on CD-ROM

Code Consultant on CD-ROM

Structural CD Series — Seismic

Beam Calculator on CD-ROM

Training videos

1999 National Building Code

WoodWorks: Software for Wood Design on CD ROM

Conventional Light Frame Construction Series:

    Span Tables

    Wood Structural Panel Grade Stamp Identification

    Proper Use of Sawn Lumber Grade Stamps

    Construction and Inspection of Floor Joists

    Framing of Stud Walls

    Framing of Ceiling Joists and Rafters

    Roof and Roof Structures

Code Information online *www.icbo.org*

Appendix A

Codes and Standards

Building codes and industry standards are closely related. Both are based on the consensus of the industry, design professionals, fire prevention agencies and many others. There are three model building codes used in the United States: the *Uniform Building Code* (UBC), the *Standard Building Code* (SBC), and the *Basic Building Code* (BOCA). Generally, the UBC is used in the western states, the SBC is popular in the southern states and the BOCA is favored in the eastern states. These documents are the best construction textbooks available. The three basic codes have largely been combined into what is now published as the 2000 IBC and the 2000 IRC.

The 1995 *CABO One and Two Family Dwelling Code* is a joint effort to standardize the requirements of all the building codes for one and two family residential buildings. The CABO code was jointly sponsored by the Council of American Building Officials (CABO), International Council of Building Officials (ICBO), Building Officials and Code Administrators (BOCA), and Southern Building Code Congress, International (SBCCI).

The model codes are published for adoption by state or local governments by reference only. Jurisdictions adopting them may make necessary additions, deletions and amendments in their adoptive documents.

Many industrial organizations associated with building construction have established standards. Some of them are the American Institute of Timber Construction (AITC), Western Wood Products Association (WWPA), and American Plywood Association (APA). There are many others. These organizations have subdivisions and committees that research and report on specialty work like wood trusses, glu-lam timbers and engineered wood products.

In this appendix I'll describe some of these authorities, their scope of interest and some of their publications.

Building Codes

Uniform Building Code

The *Uniform Building Code*, 1997 edition, is published by International Conference of Building Officials (ICBO). Their address is 5360 Workman Mill Road, Whittier, CA 90601. This code contains three volumes:

▌ Volume 1: Fire and life safety, administrative and field inspection

▌ Volume 2: Structural engineering design

▌ Volume 3: Materials, testing, installation and standards

Like the other major building codes, the *Uniform Building Code* sets minimum standards for building construction. The purpose is to provide for the safety of occupants, safeguard property and protect the public welfare. The code is continuously being upgraded as new information and products are adopted. The 1997 Uniform Building Code is the last edition published. It has been replaced by the 2000 International Building Code and the 2000 International Residential Code and its supplements. These codes have been adopted by many, but not all, building departments in the U.S.

The format of the 1997 *Uniform Building Code* has many changes from previous editions. Basic engineering moved from Chapter 23 to Chapter 16. Wood construction moved to Chapter 23 from Chapter 26. New subsections in Chapter 23 relating to wood construction are:

2301	Scope
2302	Definition and Symbols
2303	General
2304	Stresses
2305	Identification
2306	Horizontal Member Design
2307	Column Design
2308	Flexural and Axial Loading Combined
2309	Compress at Angle to Grain
2310	Tension Design
2311	Timber Connectors and Fasteners
2312	Structural Glued-Laminated Timber Design
2313	Design of Glued Built-up Members
2314	Wood Shear Walls and Diaphragms
2315	Fiberboard Sheathing Diaphragms
2316	Wood Combined with Masonry or Concrete
2317	Decay and Termite Protection
2318	Wall Framing

2319   Floor Framing

2320   Exterior Wall Coverings

2321   Interior Paneling

2322   Sheathing

2323   Mechanically Laminated Floors and Decks

2324   Post-Beam Connections

2325   Fastenings

2326   Conventional Light-Frame Construction Provisions

The National Evaluation Service is a subsidiary corporation of the ICBO. When new products appear, such as engineered wood I-beams, they must be tested in an approved laboratory and evaluated. When approved by the National Evaluation Service, the product is assigned a National Evaluation Report (NER) number which is published in the ICBO's *Building Standards* magazine. This booklet is published bimonthly and is kept on file in the member building departments. This helps keep plan checkers and building inspectors aware of new products approved for use in construction. The NER number of each new product must be renewed annually.

Typical products with NER numbers include glued-laminated beams and decking, fiber decking, oriented strand boards, particleboard, wood I-beams, truss joists, prefabricated wood trusses and laminated veneer lumber. It's risky to use any construction product which doesn't have an active NER number. If you've used a product that isn't approved, the building inspector can order you to tear it out of the building.

Standard Building Code

The *Standard Building Code* was published by Southern Building Code Congress International. Their address is 900 Montclair Road, Birmingham, AL 35213. Its chapters relating to wood construction include:

1) Administration

2) Definitions

3) Reference Standards

4) Clarification of Buildings by Occupancy

5) Special Occupancy Requirements

6) Classification of Buildings by Construction

12) Minimum Design Loads

13) Foundations

17) Wood Construction

18) Lathing and Plastering

21) Safeguards During Construction

32) Installation of Roof Framing

The *Standard Building Code* generally follows the National Design Specifications for Wood Construction, 1999 edition. These standards are published by the National Forest & Paper Association (formerly called the National Forest Products Association). The NDS specifications, generally accepted as conforming to good engineering practice, include:

- The minimum lumber grades for light frame construction
- Design values for joists and rafters
- Tables for sizing load and non-load-bearing studs
- Allowable roof/floor spans for wood and structural panel sheathing
- Allowable wood structural panel spans for combination subfloor/underlayment
- Allowable particleboard spans for combination subfloor/underlayment
- Maximum stud spacing
- Allowable spans for wood structural panel wall sheathing
- Allowable spans for particleboard wall sheathing
- Plywood exterior wall covering
- Minimum thickness of wall sheathing
- Wall bracing
- Header design charts
- Allowable shear in vertical and horizontal diaphragms made of wood structural panels and particleboard
- Diagonally sheathed shear panels nailing schedule

Basic Building Code

The *BOCA/National Building Code/1999* is published by the Building Officials & Code Administrators International, Inc. Their headquarters is located at 4051 W. Flossmoor Road, Country Club Hills, IL 60478-5795.

BOCA was founded in 1915 for the purpose of supporting sound and progressive construction. Chapters in their model code which deal with wood frame construction include:

- Administration: Inspection, permits, demolition, architectural and engineering services
- Use or occupancy: Assembly, business, educational, storage, factory and residential buildings
- Special use and occupancy: Covered malls, high rise, atriums and garages
- General building limitations: Height and area

- Types of construction: Types 1, 2, 3, 4 and 5

- Roofs and roof structures: Weather protection, fire resistance, and insulation

- Structural loads: Design safe loads, dead loads, live loads, wind loads, snow loads, seismic loads and impact loads

- Structural tests and inspections: Design strength of materials and test safe loads

- Foundations and retaining walls: Footing depth, design, mat foundations, and wood piles

- Wood: Heavy timber, wood frame, structural panels, particleboard, fiberboard, preservatives and joist hangers

- Site work: Demolition and construction, public protection and excavation

- Existing structures: Additions, alteration and moved structures

BOCA also provides a useful guide on how to use the building code to:

- Determine the requirement for construction documents

- Find the appropriate use group classification of the building

- Select the type of construction based on the building use group and the height and area limitations

- Find the required type of construction by the building material and fire resistant rating of the building elements

- Determine the location of the building site, including separation distances from lot line and other buildings

- Find the fire resistance requirements

- Determine the special use and occupancy requirements

- Calculate the structural loads and stresses

- Determine the building service requirements

International Building Code

The three basic building codes, the UBC, BOCA, and SBC have been replaced by a single common code. The new code is published as the International Building Code and the International Residential Code. They were first published in 2000, and supplemented in 2001 and 2002. The new code merges the best aspects of each of the previous codes. There is also a central location for accumulating additions and proposals for changes to the code. Other advantages are the ability to handle governmental issues better, create universal educational programs and produce standardized products for national use.

Standards

National standards are published by many of the industrial organizations associated with wood building construction.

American Institute of Timber Construction (AITC)

This organization is located at 7012 S. Revere Parkway, Inglewood, CO 80112-3932, Phone: 303-792-9559, Fax: 303-792-0669. Their most important publication is the *Timber Construction Manual*, 3rd edition, 1985, John Wiley and Sons, Inc. This manual, written for architects, engineers, contractors, wood laminators and fabricators, deals with engineered timber buildings and other structures. Some of the subjects covered in this book are:

- Physical and mechanical properties of wood
- Fastenings and connections
- Design of structural timber
- Standard lumber sizes
- Trigonometric formulas
- Beam diagrams and formulas

Some of their other publications include:

- AITC 104 *Typical Construction Details*
- AITC 110 *Standard Appearance Grades for Glued Structural Laminated Timbers*
- AITC 112 *Standard for T&G Heavy Timber Roof Decking*
- AITC 113 *Standard for Dimensions of Structural Glued Laminated Timber*

American Plywood Association

American Plywood Association (APA) is a nonprofit trade association whose member mills produce approximately 80 percent of structural wood panels products (plywood) manufactured in the U.S. Their mailing address is P.O. Box 11706, Tacoma, WA 98411-0700, Phone: 253-565-6600, Fax: 253-565-7265. Some of their publications helpful to wood designers are:

- *Nailed Structural-Use Panel & Lumber Beams*
- *Design and Fabrication of Plywood Stressed-Skin Panels*
- *Design and Fabrication of Plywood Sandwich Panels*
- *Design and Fabrication of Glued Plywood-Lumber Beams*
- *Design and Fabrication of All-Plywood Beams*
- *Roof Sheathing Fastening Schedules for Wind Uplift*
- *Plywood Diaphragm*
- *Plywood Design Specifications*

Western Wood Products Association (WWPA)

The mailing address of this association is 1500 Yeon Building, 5622 SW Fifth Avenue, Portland, OR 97204. Their principal publication is *Wood Frame Design* for commercial and multifamily construction, which is filled with drawings and details. It covers the following subjects:

- Wood design
- Shrinkage and compression
- Connectors, failures and recommendations
- Fire stopping and draft stopping
- Sound control and fire resistance

National Forest & Paper Association

This association is located at 1111 19th Street N.W., Seventh Floor, Washington, DC 20036. The most important publications include the following:

- ANSI/NFoPA NDS-1991 *National Design Specification for Wood Construction* (dated October 16, 1992)
- *Design Values for Wood Construction* which is a supplement to the NDS.

Another book is called *Forest Service - Agriculture Handbook No. 72*. This book provides engineers, architects and others data on the physical and mechanical properties of wood. It also includes lumber stress grades and allowable stresses, fastenings, nails, screws, bolts and connectors, glued structural members, sandwich construction and preservation of wood.

Other Associations

There are many other industrial associations that relate to wood-frame construction. I'll list them in alphabetical order, including their addresses and phone numbers.

American Hardboard Association
1210 W. Northwest Highway
Palatine, IL 60067
847/934-8800

American Parquet Association
2900 First Commercial Building
Little Rock, AR 72201
501/375-5561

American National Standards Institute
11 West 42nd Street
New York, NY 10036

American Society of Testing and Materials
100 Barr Harbor Drive
West Conshohoken, PA 19428-2959

American Wood Council

PO Box 5364

Madison, WI 53705-5364

The American Wood Council (AWC) is the wood products group of the American Forest & Paper Association. AWC supports the development of sound building code practice related to wood construction. They promote state-of-the-art engineering data, technology and standards for engineers, architects, builders and building code administrators. Some of their publications are:

Publications available:

National Design Specifications for Wood Construction (1991)

Design Values for Wood Construction (1991)

Wood Structural Design Data (1986 Edition with 1992 Revisions)

Span Tables for Joists and Rafters (1993)

Design Values for Joists and Rafters (1992)

Wood Frame Construction Manual for One- and Two-Family Dwellings (1995 SBC High Wind Edition)

Wood Frame Construction – Manual and Commentary – National Edition (2001)

WCD No. 1 – *Manual for Wood Frame Construction* (1988)

WCD No. 4 – *Plank and Beam Framing for Residential Buildings* (1989)

WCD No. 5 – *Heavy Timber Construction Details* (1989)

WCD No. 6 – *Design of Wood Frame Structures for Permanence* (1988)

American Wood-Preservative Institute

PO Box 5690

Grandbury, TX 76049

The AWPA is an international technical society dedicated to exchange of information relating to the wood preserving industry. They are recognized by most users of treated wood such as building, electrical, marine and road construction.

Publications available:

Book of Standards (1995)

Wood Preserving Statistics (1993)

American Wood Preservers Institute

1945 Old Gallows Road, Suite 550

Vienna, VA 22182

703/893-4005

Builders Hardware Manufacturers Association

355 Lexington Avenue, 17th Floor

New York, NY 10017

212/661-4261

California Redwood Association

405 Enfrente Drive, Suite 200

Novato, CA 94949

888/225-7339

415/382/0662

415/382-8531 (Fax)

info@calredwood.org

> ***Publications available:***
>
> *Redwood Design-A-Deck Kit*
>
> *Redwood Specifier* (for builders and architects)

Federal Housing Administration

Department of Housing and Urban Development

451 7th Street SW

Washington, DC 20410

202/708-1122

> ***Publication available:***
>
> *Minimum Property Standards for One and Two Family Dwellings*

International Code Council

5203 Leesburg Pike, Suite 708

Falls Church, VA 22041

National Association of Home Builders of the US

15th and M Streets, NW

Washington, DC 20005

202/822-0200

National Forest Products Association

1250 Connecticut Avenue NW, Suite 200

Washington, DC 20036

202/463-2700

National Frame Builders Association

4840 W. 15th Street, Suite 1000
Lawrence, KS 66049-3876
913/843-2444
913/843-7555 (Fax)

The NFBA represents post-frame construction, manufacturing, code and design professionals.

Publications available:

Post-Frame Building Design

Post-Frame Construction Comes of Age

Post-Frame: Code Conforming Construction for Commercial, Industrial and Institutional Applications (a video for code professionals)

National Oak Flooring Manufacturers Association

PO Box 3009
Memphis, TN 38173-0009
901/526-5016

The NOFMA represents producers of oak flooring in the United States. It sets up standards for the manufacturing and grading, and regularly inspects the member mills for compliance with their grading rules. The Oak Flooring Institute is an affiliate of NOFMA.

Publications available:

Installing Hardwood Flooring

Finishing Hardwood Flooring

Behavior of Flooring Shrinkage

Cracks in Hardwood Flooring

Wood Floor Care Guide

National Particleboard Association

18928 Premiere Court
Gaithersburg, MD 20879
301/670-0604
301/840-1252 (Fax)

The NPA is a trade association of manufacturers of particleboard (PB) and medium density fiberboard (MDF). The NPA publishes technical information on warp. Screwholding, sawing, joint design, laminating and floor underlayment for architects, engineers and builders

Publications available:

Buyers and Specifiers Guide to Particleboard and MDF

Particleboard From Start to Finish

MDF From Start to Finish

Various American National Standards on PB and MDF

Northeastern Lumber Manufacturers Association

272 Tuttle Road

Cumberland Center, ME 04021

207/829-6901

Publications available:

Standard Grading Rules for NeLMA and NSLB

Siding Installation Chart

Eastern White Pine and Structural Grades brochures

Southern Forest Products Association

PO Box 641700

Kenner, LA 70064

504/443-4464

Publications available:

Exterior Wood in the South, Selection Application and Finishes

Guide to Southern Pine Siding, Patterns, Installation and Maintenance

Maximum Spans – Southern Pine Joists and Rafters

Southern Pine Use Guide

Pressure Treated Southern Pine

Southern Pine Floor Trusses

Non-residential Case Studies – Uses of Southern Pine Engineered Wood Systems

Engineered Wood Systems

Construction Guide – Southern Pine

Southwestern Lumber Manufacturers Association

PO Box 1788

Forest Park, GA 30051-1788

404/361-1445

404/361-5963 (Fax)

Publications available:

The Hardwood Handbook

Western Red Cedar Association

PO Box 2786

New Brighton, MN 55112

612/633-4334

The WRCA represents firms who produce, manufacture, and treat Western Red Cedar poles for electrical utility industry. Poles vary from 30 feet to 120 feet and are sold in round form only.

Western Red Cedar Lumber Association
100/555 Burrard Street
Vancouver, BC Canada V78 1S7
604/684-0266
604/682-8641 (Fax)

PO Box 2888
Naperville, IL 60567-2888
708/369-2828
708/369-8651 (Fax)

No. 203 - 457 Main Street
Farmingdale, NY 11735
516/643-9225
516/643-7252 (Fax)

> ***Publications available:***
> *Specifications for Cedar Siding*
> *Installing Cedar Siding*

Western Wood Products Association
1500 Yeon Building
522 SW Fifth Avenue, Suite 500
Portland, OR 97204-2122
503/224-3930

> ***Publications available:***
> *Western Wood Use Book*
> *Product Use Manual*
> *Wood Frame Design*
> *Lumber Basic CD*
> *Span Tables online*
> *Lumber mold info on line*

Appendix B

Glossary

Acceleration (earthquake) - The rate of ground movement as compared with the acceleration of a falling object.

Accepted engineering practice - Design and construction that conforms to accepted principles, tests or standards of nationally-recognized technical and scientific authorities.

Acrylic (wood) - A wood-plastic composite produced by impregnating wood with an acrylic monomer polymerized by ionizing radiation or other techniques.

A.D. - Air dried.

Addenda - Documents issued before the bid opening which clarify, correct or change bidding documents or the contract documents.

Adhesive - A substance capable of holding materials together by surface attachment.

Aftershock (earthquake) - An earthquake of less intensity than the initial earthquake.

Agreement - The document signed by the owner and contractor covering the work to be performed.

AHA - American Hardwood Association.

Air-dried (A.D.) lumber - Lumber that has been stored in an open area to dry naturally. Also see *Seasoning*.

AITC - American Institute of Timber Construction.

Allowable span - The greatest horizontal distance permitted between supports.

Allowable stress - The amount of force per unit area permitted in a structural member.

Allowable stress increase - A percentage increase in the allowable stress based on the length of time that the load acts on the member.

Anchor base - An embedded anchor that includes a seat in which a framing member can rest and be fastened.

Anchor bolts - Steel bolts embedded in concrete which hold a building or structure to the foundation.

Annual growth rings - Layers of wood growth added to a tree during a single growing season.

ANSI - American National Standard Institute

APA - American Plywood Association, which represents most of the plywood manufacturers for the purpose of research, quality and promotion.

Architect - A person licensed by the state charged with the design and specifications of a building.

ASTM - American Society of Testing and Materials.

AWPA - American Wood Preservative Association

Axial force - A push (compression) or pull (tension) acting along the length of a member, usually measured in pounds.

Axial stress - The load or force divided by the cross-sectional area of the member, usually expressed in pounds per square inch (psi).

Back (plywood) - The back veneer of a plywood panel which is normally of lower quality than the front veneer.

Backshore - A shore placed under a concrete slab or beam after the formwork and original shores have been removed.

Backstay - See *Brace.*

Balloon framing - A framing system whose studs are continuous from the bottom to the top of a building.

B&S - Beams and stringers.

Barge rafter - The outermost rafter on a gable roof.

Base shear (earthquake) - The total horizontal seismic force exerted at the top of a foundation.

Basic wind speed - The probable fastest wind speed measured 33 feet above the ground in a flat open area.

Batter - An inclination from the vertical.

Batterboards - Elevated horizontal boards set at the corner of a building used to establish the locations of the corners and elevation of the building foundation wall.

Beam - A horizontal load-bearing structural member.

Beam bottom - Soffit or bottom form for a concrete beam.

Bearing - An angular deviation of a line compared with the north-south line. This angle is measured in degrees, minutes, and seconds.

Bearing wall - A wall that supports a load in addition to its own weight.

Bench mark - A permanent landmark that has a known position and elevation. When used with a surveyor's level, a bench mark is the primary reference point for finding the elevation of other points.

Bending moment - A measure of bending effect due to a load acting on a member which is usually measured in foot pounds, inch pounds, or inch kips.

Bending stress - the force per square inch of area acting at a point along the length of a member resulting from the bending moment applied at that point, usually applied in psi.

BHMA - Builders Hardware Manufacturers Association

Bid - The written proposal submitted by a bidder stating the prices for the work to be performed.

Blind nailing - Driving nails so that nail heads are concealed.

Blocked diaphragm - A diaphragm in which all sheathing edges are supported by framing members or blocking.

Blocking - Small wood pieces installed between studs, joists or rafters to prevent buckling.

Board foot - A unit of measure equivalent to a board 1 foot square and 1 inch thick.

Boards - Lumber 2 or more inches wide and not more than 1½ inches thick.

Borer holes - Wood voids made by grubs, worms and wood-boring insects.

Bottom plate - See *Sill plate*.

Boundary element (diaphragm) - The perimeter or edge of a plywood sheathed roof, floor or wall.

Bow - A lumber distortion parallel to the grain.

Box beam - A built-up beam with solid wood flanges and a plywood or wood-base panel web.

Box-out - An opening or pocket formed in concrete.

Brace - A load-bearing member installed diagonally.

Bridging (or cross-bridging) - Diagonal bracing placed between floor joists and rafters to keep them from buckling.

Buck - The framing around an opening in a wall.

Builder's level - A surveying instrument to control horizontal planes.

Building line - A line established by law as the minimum distance from the street line to the face of the building. Also called a *setback line*.

Building paper - Heavy paper used to waterproof walls and roofs.

Building permit - A document issued by the building department certifying that plans have been approved for construction.

Built-up member - A single structural wood member made from several pieces fastened together.

Built-up roofing - Two or more layers of roofing consisting of a base sheet, felts, cap sheet, mineral aggregate, smooth coating or similar material.

Bulkhead - A partition built into wall forms to terminate each pour of concrete.

Butt joint - A straight wood joint with an interface perpendicular to the grain.

Camber - A predetermined curve set into a steel beam during fabrication to make up for the sag that will occur when the beam is loaded.

Cants - Logs cut on two or four sides.

Cant strip - See *Chamfer*.

Catwalk - A narrow elevated walkway.

Ceiling joist - A horizontal structural member that spans the tops of the wall and supports the finish ceiling.

Centering - Temporary supports placed under arches, shells and space structures that are lowered as a unit to prevent destructive stresses on the structure.

Chalk line (snap line) - A spool-wound string encased in a chalk-filled container that is pulled taut across a surface, lifted and snapped directly downward so that it leaves a straight chalk mark.

Chamfer - A beveled edge formed in concrete by a triangular strip of wood (chamfer strip) placed in the form corner.

Change order - A written order to the contractor signed by the owner authorizing him to add, delete, or revise the work.

Check (lumber) - A lengthwise separation of wood that usually extends across or through the annual growth rings.

Chord (truss) - The top and bottom members of a truss.

Cleanout - An opening at the bottom of forms that allows access for removing refuse.

Clear (hardwood) - Wood free of defects except for burls, streaks, and pinworm holes.

Clear span - The horizontal distance between the interior edges of the supports for a beam or truss.

CLR - Clear.

Collar beam - A horizontal tie beam in a gable roof connecting two opposite rafters near the ridge.

Column (post) - A vertical load-bearing structural member.

Combined stress - A combination of axial and bending stresses acting on a member simultaneously.

Common nail - A steel wire nail.

Composite panel - A structural panel made of layers of veneer and wood-based materials.

Compression - A force that tends to crush a wood member.

Concentrated load - A load centered at a given point.

Concentrically braced frame (earthquake) - A braced frame in which the members are subjected primarily to axial forces.

Conifer trees - Cone-bearing trees from which softwoods are derived.

Construction documents - All the written, graphic and pictorial documents describing the design, location and physical characteristics of a building necessary for obtaining a building permit.

Conventional light-frame construction - Framing consisting of wood studs, joists, rafters and flooring. Commonly called *Stick Framing*.

Corner brace - A member installed diagonally at the corners of a wood frame building.

Course - The direction of a line. It's given with respect to north or south. The same line may be described as 45 degrees east of north (N 45° E) or 45 degrees west of south (S 45° W).

Cricket - A small roof structure installed to provide slope away from wall or chimney.

Cripple - Short vertical framing member installed above and below windows; any part of a vertical frame which is cut less than full length.

Cripple wall - A wall installed between the first floor and the foundation.

Cross-bridging - See *Bridging*.

Cross section (beam) - A section taken through a member perpendicular to its length.

Crown - An upward bow in a horizontal structural member.

Crowning - Installing a horizontal member with its crowned edge up.

---

d - An abbreviation for "penny" in designating nail size.

Dead load - The permanent weight of a building structure, including equipment.

Decay - The deterioration of wood due to fungi.

Decay fungus - A living organism that sends minute threads of "hyphae" through the damp wood for food.

Decenter - To lower or remove centering or shoring.

Decking - Sheathing material used for a deck or slab soffit forms.

Deflection (beam or truss) - The amount of sag in a horizontal structural member, usually expressed as a ratio of the amount of deflection to the span of the beam.

Deformation - Any change in shape, including shortening, lengthening, twisting, buckling or expanding.

Delamination (plywood) - The separation between plies, normally due to moisture.

Dense select structural - A high-quality lumber, relatively free of characteristics which impair its strength or stiffness.

Depth - The board dimension measured parallel to the direction of the principal load on the member.

Deputy inspector (special inspector) - A specially-approved building inspector.

Design earthquake - This design standard has a 90 percent probability of not being exceeded within 50 years.

Design load - The total load that a structural member is designed to support.

Diaphragm - A horizontal or nearly horizontal system designed to transmit lateral forces to the vertical structural members of a building.

Diaphragm Boundary - See *Boundary element*

Diaphragm chord - The outer edges of a horizontal or vertical diaphragm.

Diaphragm strut - A compression or tension member that transfers horizontal seismic loads to a diaphragm.

Diaphragm (vertical) - See *Shear wall*.

Dimension lumber (standard dressed lumber) - Lumber 2 to 5 inches thick and up to 12 inches wide, including joists, rafters, studs, planks, posts and small timbers.

Distance of a course - The length of the course given in feet and decimal parts of a foot.

Distress - Signs of failure due to overstress.

Domain - See *Tributary area*.

Doubling - Nailing two structural members together to form one.

Douglas fir lumber - A common species of tree grown in the western United States.

Dowel - A pin used to join two pieces of wood together.

Drift - The horizontal movement of a building caused by an earthquake.

Dry rot - Deterioration of wood caused by fermentation and chemical breakdown when it is attacked by fungus.

Drywall - See *Gypsum wallboard*.

Drywall shear wall - An earthquake- or wind-resistant wall clad with gypsum wallboard sheathing.

Dumpy level - A surveyor's instrument used to control horizontal planes.

Earthquake magnitude - The effects of an earthquake measured with the Mercalli Scale.

Easement - An interest in the land of another person. This may entitle the owner of the interest certain use of the land, such as the right to cross over the land (affirmative easement). It may prohibit the owner from doing something to the land, such as constructing a tall building (negative easement). Caution must be taken within an easement for underground utilities, such as gas transmission pipelines or electrical conduits.

Eaves - The edges of the roof that extend beyond the exterior walls of the building.

Edge nailing (plywood sheathing) - A series of nails along each edge of a plywood panel.

Elasticity - See *Modulus of elasticity*.

End nailing - Nails driven into the end of a board.

Engineered wood - A specially-designed structural member or assembly that is usually built off-site.

Equilibrium moisture content - The moisture content of wood that is in balance with the relative humidity.

Exposure (wind) - A description of the terrain surrounding a building and the highest wind velocity with regard to wind exposure.

Fabricated structural timber - An engineered wood member, including sawn lumber, glued-laminated timber and mechanically-laminated lumber.

Face (plywood) - The side of a plywood panel that is of higher veneer quality when front and back are of different veneer grades.

Factor of safety - The allowable unit stress based on the judgment of a competent authority.

Falsework - A temporary structure erected to support work in progress such as shores, or vertical posts supporting formwork.

Fascia - A board attached to the ends of the rafters to create a finished edge around the roof.

Fascia rafter - The outermost rafter at the end of a gable roof.

Fastest mile wind speed - The highest sustained average wind speed on a mile-long sample of air passing a fixed point.

Fault (earthquake) - A zone of weakness in the earth's crust allowing movement between adjacent crust blocks.

FBM - Feet board measure.

Fiberboard - A construction material made of wood or other plant fiber compressed into large sheets.

Fiber reinforced gypsum panel - A construction material made of gypsum slurry and plant fibers formed into large sheets.

Fink truss - A type of open triangular-shaped truss.

Fire blocking - See *Fire stop*.

Fire-rated - A construction material that is tested and shown to be fire-resistant for a given period of time.

Fire-resistive construction - Construction in which the structural frame is protected from fire by covering it with plaster or gypsum wallboard.

Fire-resistive rating (fire resistance) - The time in hours (or fractions thereof) during which a material will withstand exposure to fire based on ASTM testing procedures.

Fire-retardant wood - Any wood product pressure-treated with chemicals that shows a flame-spread index of 25 or less when tested according to ASTM E-84.

Fire stop (blocking) - Horizontal 2-by blocks installed between studs to prevent spread of fire and smoke.

Flame spread - The rate at which a flame spreads over a surface.

Flange (built-up beam) - The horizontal members of an I-beam or box beam.

Flexural strength - Resistance to bending stresses.

Floor girder - A beam that supports floor joists.

Force (earthquake) - The acceleration of a building resulting from sudden earth movement.

Force - A push or pull exerted by one object on another.

Force diagram - A graphic solution of forces as they interact within a structural system.

Formwork - The total system built to contain freshly-placed concrete, including sheathing, supporting members, hardware and bracing.

Foundation - The brick or concrete support wall a house or building sits on.

Framing hardware - Metal connecting devices used in wood-frame construction.

FRT plywood - Fire-retardant plywood panels used for roof decks.

Full size lumber (sawn lumber) - Undressed or rough lumber.

Fungi (wood) - Microscopic plants that live in damp wood and cause deterioration and mold.

Furring - Wood strips installed to provide a base for finish material.

Gable - The end wall that extends from the top plate to the ridge.

Gang nails - Light-gauge metal plates used as connectors for wood members.

Ganged forms - Prefabricated form panels joined to make a larger unit for efficiency in erecting, stripping and reuse.

Gin pole - A nearly vertical wood post used to hoist heavy material and equipment.

Girder - A major horizontal structural member that supports secondary beams, joists or rafters.

Girt - A horizontal member used to support wall siding.

Glued-laminated (glu-lam) timbers - Structural members made of wood, plywood or both, bonded together with adhesive.

Grade-marked lumber - Lumber that has been inspected and stamped showing the species and quality of the wood.

Green lumber - Freshly sawn, unseasoned or undried wood.

Gusset - A small piece of wood, plywood or metal attached to the corners or intersections of a frame to add strength and stiffness.

Gypsum wallboard - A building material made with a gypsum core covered with paper. It is applied to interior walls. Also called *gypsum board, sheetrock* or *drywall.*

Hardwood - Heavy and close-grained lumber from broad-leafed deciduous trees.

Header - A horizontal beam installed perpendicular to joists, or a lintel over a door or window.

Header joist (ribbon or band joist) - A horizontal member butted against the ends of floor joists.

Heartwood (summerwood) - Wood between the sapwood and the center core (pith) of a tree.

Heartwood decay - Only occurs in living trees.

Heavy timber construction - A fire-resistant wood frame consisting of thick decks and large posts and beams.

Heel (truss) - The part of a roof truss where the top and bottom chords intersect.

Hip roof - A roof that has inclined planes from all four sides of a building.

Hold-down connectors - Steel anchor straps used to bolt a shear wall to the foundation.

Horizontal seismic force - The reaction of a building or structure to the movement of the ground during an earthquake.

HPVA - Hardwood Plywood Veneer Association

Hubs - Wood stakes, usually 12 inches long, made of 2 x 2 hardwood, pointed at one end and painted white at the top.

ICC - International Code Council

Identification index (plywood) - See *Span rating.*

In-line joists - Floor joists butted end-to-end and fastened together with splice pieces nailed to both sides of the joint.

Intermediate nailing (sheathing) - A series of nails within the interior of plywood panels.

Jack - A mechanical device used to adjust the elevation of forms or form supports.

Jack rafter - A rafter that spans from a wall to a hip or valley rafter.

Jack shore - A telescoping adjustable single-post metal shore.

Jack truss - A truss that supports another truss, replacing a post.

Joists - Horizontal structural members that support a floor or ceiling.

Journeyman - A tradesman with the experience required to complete any task without supervision.

K.D. - Knocked down.

Keel - An oil crayon used to mark the locations of framing members.

Kerf - A notch or cut in a beam.

Kiln - A chamber whose controlled environment rapidly dries lumber.

Kiln-dried lumber - Wood seasoned in a special chamber using artificial heat.

King studs - The studs of either side of a window or door, which support the header.

Kip - A unit of force representing a thousand pounds.

Knee brace - A corner brace placed at an angle.

L-head - The top of a shore that has a braced horizontal member projecting on one side forming an inverted L-shaped assembly.

L-shore - A shore with an L-head.

Lagging - Heavy sheathing used for underground work to temporarily support earthen walls.

Laminated purlins - Structural roof members made of layers of wood strips.

Laminating - The process of bonding wood laminations together with adhesive.

Lap joint - A connection made by placing two pieces of material side by side and joining them together with nails, bolts or glue.

Lateral brace (support) - A member installed at right angles to a chord or web members of trusses for alignment and support.

Lateral load (force) - A side-to-side force acting on a structure.

Layer (plywood) - A single veneer ply or two or more plies laminated with a parallel grain direction.

Layout bar - A steel bar with small cross bars used for laying out stud locations.

Lead and tack (L&T) - A means of marking a survey point in concrete by chiseling, filling with lead, and setting a brass tack.

Ledger - A horizontal member attached to a wall or girder to support joists or rafters.

Legal description - A written description that allows an owner to establish, maintain, and transfer the right to occupancy and use. The description may be by government subdivision, metes and bounds, or reference to a tract map.

Let-in brace - A diagonal brace that is set into a notched stud so it fits flush with the stud.

Licensed surveyor (L.S.) - A surveyor who is registered to practice land surveying under the state regulations.

Life safety design - Structural design formulated to prevent the collapse of a building due to earthquake or fire.

Lintel (header) - A horizontal member installed over an opening in a wall to support the wall construction above.

Live load - Any load which is not permanent, such as people and temporary construction loads.

Lot and tract number - Number that identifies the lot and tract of a subdivision.

Lumber - Wood that has been sawed, planed and cross-cut to length.

Machine stress-rated lumber (MSR) - Mechanically-graded lumber not more than 2 inches thick and at least 2 inches wide. Also called machine-evaluated lumber (MEL).

MBF - One thousand board feet.

MBM - One thousand (feet) board measure.

Mechanically-laminated - A laminated wood structural member held together with mechanical fasteners.

Metes and bounds - A method of naming irregular shaped parcels. The land is described by running around the boundary by courses, distances and fixed monuments at the corners of angles. Metes mean measurements and bounds mean boundaries.

Modulus of elasticity - The modulus of elasticity of a given material is determined experimentally by placing the material in tension (or compression) with the application of a known force. Then the unit deformation (elongation or contraction) suffered by the material is measured and defined as the ratio between the unit stress applied and the unit deformation suffered.

Modulus of rupture - The maximum bending stress.

Moisture content (wood) - The weight of water in wood divided by its dry weight.

Moment of inertia (cross section of a beam) - A measurement of a structural member's ability to resist changing shape. It indicated the member's strength.

Monuments - Land surveyors locate the boundary of parcels by fixed marks on the ground called monuments.

Mud sill - A wood member (usually treated or redwood) bolted to the top of a foundation to support floor joists. Also called a *sill plate* or *sole plate*.

Multi-tier shoring - Single-post shores used in two or more tiers to increase the height of the shoring platform.

NDS - The National Design Specifications for Wood.

NELMA - Northeastern Lumber Manufacturers Association.

Neutral axis (cross section of a beam) - A point within a beam where there is neither tension or compression stress.

NFPA - National Forest Products Association.

Nominal size (lumber) - The rough size of lumber before it is finished and surfaced. A dry *nominal* 2 x 4 is about $1^1/_2$ by $3^1/_2$ inches.

Nominal thickness (plywood) - The full designated thickness of plywood before sanding.

Nominal span - The horizontal distance between the edges of the supports of a beam or truss.

Non-bearing wall - A wall or partition that only carries its own weight.

Occupancy - The purpose for which a building is to be used.

On center (OC) - The distance between the centers of adjacent repetitive structural members.

Outrigger - Lumber that extends beyond a rake to support fascia rafters.

Oriented Strand Board (OSB) - A wood structural panel that is a mat-formed product composed of thin rectangular wood strands or wafers arranged in oriented layers.

Parapet wall - Part of an exterior wall above the roof line.

Particleboard - A mat-formed panel made of wood particles or a combination of wood particles and wood fibers bonded together with synthetic resins.

Partition - An interior wall that subdivides building spaces.

Peck - Channeled or pitted areas or pockets of fungus that only occurs in live trees.

Pier - A masonry or concrete column used to support a beam.

Pier block - A preformed concrete footing that supports a post.

Pitch pocket - An opening in a wood member containing liquid pitch.

Pith - The soft center core of a tree.

Plank - 1) A wide piece of sawed lumber, usually $1^1/2$ to $4^1/2$ inches thick and 6 or more inches wide. 2) For flooring, boards thicker than $3/4$ inch, usually made of 2 inch nominal thickness or $1^1/2$ inch actual thickness lumber.

Platform framing - A wood structural system in which the studs of each story extend from the sill plate to the top plate of each story.

Plumb - True and level on a vertical plane.

Plumb bob - A metal weight suspended from a cord used to establish a vertical line.

Ply - A single veneer lamina in a glued plywood panel. Also, the number of thicknesses of veneer in a plywood panel or laminated member.

Plyform - Plywood designed for high re-use concrete forming; may be mill-oiled.

Pneumatically-driven fasteners - Air-driven nails, staples or spikes.

Post (column) - A vertical load-bearing structural member.

Property corner - A geographic point on the surface of the earth that controls the property line.

Purlin - A horizontal member that acts as a beam and supports common rafters or ceiling joists.

Push stick - A carpenter's device used to plumb a wall during erection.

Racking - A twisting movement that can distort a framework.

Rafter - A sloping beam from the ridge of a roof to the eaves, supporting the roof.

Rake - The sloping edge of a gable roof.

Reaction - The load transmitted from a beam or truss to a support.

Registered design professional - Any architect or engineer registered or licensed to design a project in a given state according to the state's professional registration laws.

Reshore (reshoring) - To remove the form and shores together, and replace each post immediately with a new post wedged into place to support the concrete.

Retrofit - To add additional bracing, anchoring or any improvement to a completed structure.

Ridge board - A horizontal member at the ridge of a roof that the tops of rafters are framed into.

Rim joist - The outermost joint on a floor, directly over the exterior walls. Rims also surround an opening for stairs or any other opening in a floor.

Riser - The vertical part of a step in a flight of stairs.

Sapwood - The portion of a tree between the bark and the inner heartwood that conducts water from the roots to the leaves.

Sawn - Wood that is sawn from one board, instead of laminated or built-up.

Saw-sized lumber - Lumber that's uniformly sawn to the net size for surfaced lumber.

Scab - A small piece of wood fastened to two formwork members to secure a butt joint.

Scaffolding - An elevated platform erected to support workers, tools and materials.

Scissor truss - A steeply-sloped truss with a bottom chord that angles up in the middle to form a sloped ceiling.

Screw jack - A threaded steel rod at the top or bottom of a shore used to adjust the length of the shore.

Seasoning (wood) - Drying lumber by exposure to air and sun or by kiln.

Section modulus - A property of the shape of a structural member that indicates its strength; the moment of inertia divided by the distance from the neutral axis to the extreme fiber of the section.

Seismic load - An assumed lateral load, caused by an earthquake, acting in a horizontal direction on a structural frame.

Select structural lumber - High-quality lumber, free of characteristics which impair strength or stiffness.

Setback lines - A line established by law as the minimum distance from the street line to the face of the building. Also called a *building line*.

Shear load - Side-to-side force(s) acting on a structure.

Shear stress - The stress that tends to keep two adjoining planes of a body from sliding on each other when two equal parallel forces act on them in opposite directions.

Shear wall - A wall designed and constructed to provide shear strength.

Sheathing - The structural covering applied to the outside surface of wall or roof frame. Also called *sheeting*.

Sheetrock - See *Gypsum wallboard*.

Shim - A long narrow repair of wood used as a wedge to level or plumb a structural element.

Shop drawing - A drawing, diagram, illustration, schedule and other data prepared by a contractor, manufacturer, fabricator, supplier or distributor to illustrate some portion of the work.

Shore - A temporary vertical or inclined member that supports formwork and fresh concrete until the structure has developed full strength.

Siding - The finish covering applied to the outer side of the exterior walls of a frame building.

Sill plate - The lowest horizontal member of a wall frame bolted to the foundation. Also called a *sole plate* or *mud sill*.

Sleepers - Treated wood nailers attached to a concrete slab providing a nailing base for flooring.

Snapping - Laying out the floor plan on the slab with a chalk line.

Snow load - The load on a building resulting from the accumulation of snow.

Soffit - 1) The underside of a structural member of a building, such as a beam. 2) A covering over the space under the eaves, or the space between a cabinet and the ceiling.

Softwood - Lumber from trees with needle or scale-like leaves.

Soldier - Vertical wales used to strengthen and align forms.

Sole plate - See *Sill plate*.

Span - The horizontal distance between supports.

Span rating (plywood) - A pair of numbers stamped on plywood sheathing to indicate its span capabilities over roofs and floors. Also called an *identification index*.

Specifications - Technical descriptions of a project regarding materials, equipment, construction systems, standards and workmanship.

Split - The separation of wood fibers caused by external forces.

Station - An increment of 100 feet along a transit line.

Stress-grade lumber - Lumber that has been assigned working stress.

Strike - To lower or remove formwork or centering.

Stringer - A beam that supports a floor or deck sheathing.

Stripping - The disassembling of forming and shoring, usually for reuse.

Strips - Boards less than 6 inches wide.

Structural lumber - Lumber that is more than 2 inches thick, used where working stresses are required.

Strut - A horizontal or inclined compression member.

Stud - 1) In concrete forms: a vertical wood or metal member that supports the sheathing. 2) In walls: a vertical wood or metal member that supports sheathing, joists and rafters.

Subcontractor - An individual, firm or corporation having a direct contract with the general contractor or with any other subcontractor for the performance of part of a project.

Sub-diaphragm - A portion of a larger diaphragm designed to transfer forces to the main diaphragm.

Subdivision - A tract of land divided into several parcels offered for sale.

Subfloor - Sheathing fastened to floor joists to support finish flooring.

Survey markers - Pipes, lead and tack, wood hubs, and brass monuments.

Sway brace - A diagonal brace used to resist wind or other lateral forces. Also see *X-brace*.

T&G - Tongue-and-groove.

T-head - The top of a shore with a braced horizontal member projecting on two sides, forming a T-shaped assembly.

Teco nail - A nail used to install hangers.

Tell-tale - Any device designed to indicate the movement of formwork.

Template - A thin plate or board frame used as a guide in positioning or spacing of form parts, reinforcement, anchors, etc.

Tensile strength - Resistance to tension forces.

Tensile stress - A pulling force over a unit area, such as a square inch. Also called *intensity of stress, stress per unit of cross-sectional area* or *unit stress.*

Tension - The force exerted on a structural member that has the effect of either pulling apart or elongating the structural member in question.

Tie beam - A beam that ties roof rafters together.

Tie wire - Iron wires used to hold opposing wall forms in position.

Timber - Lumber that's 5 or more inches in the least dimension, including beams, stringers, posts, sills, girders and purlins.

Toenail - A nail sent into a board at about a 60-degree angle, an inch or so from its edge.

Tongue-and-groove - Material with one edge shaped with a narrow projection that fits into a corresponding groove in another piece, locking them together.

Top chord - An inclined or horizontal that established the upper edge of a truss, usually carrying compression and bending stress.

Top plates - Horizontal members installed over the tops of studs supporting rafters and joists.

Touch-sanding (plywood) - A sizing operation consisting of a light surface sanding in a sander.

Transit line - A base line used to control the horizontal position of key items in construction.

Tributary area (domain) - The roof, floor or wall area that contributes to the load on a structural member.

Trimmer - The stud that's nailed to the king stud and runs up under the header in framing rough openings.

Truss - A shop-fabricated frame used as a roof support.

Ultimate stress - The maximum stress that a material can stand before it breaks apart.

Underlayment - Sheathing material used as a base for finish flooring or carpet.

Underpinning (wood) - Studs or posts installed on a foundation supporting the first floor.

Uniform load - A load that is equally distributed over a given length of a beam, and is usually expressed as pounds per linear foot (plf).

Unit stress - The amount of stress on 1 square inch of a material.

United States Coast and Geodetic Survey (U.S.G.S) - An agency of the federal government that surveys and sets monuments of the U.S. Rectangular System.

Uplift - A force or forces acting to lift a structure.

Veneer - A thin outer layer of a select wood bonded on to a less-attractive, less-expensive wood, used in paneling and furniture.

Wale (waler) - A horizontal member used to align and brace studs on concrete forms.

Wane - A lumber defect located near the edge or corner of a board caused by the presence of bark.

Web - A diagonal or vertical member that joins the top and bottom chords of a roof truss.

Western platform framing - A wood framing system in which studs stop at each floor level.

Wood rays - Flat bands that radiate from the core of a tree to the bark.

Wood shear panel - A wood floor, roof or wall frame sheathed to act as a shear wall or diaphragm.

Work - The construction required under the contract documents.

Working stress - The unit stress that experiment has shown to be safe in a material, while maintaining a proper degree of security against structural failure.

WWPA - Western Wood Products Association.

X-brace - A paired set of sway braces.

Yard lumber - Wood members used for common construction.

Yoke - A tie or clamping device place around column forms or over the top of wall or footing forms to keep them from spreading due to lateral concrete pressure.

Index

Practical References for Builders

2000 *International Residential Code*

Replacing the *CABO One- and Two-Family Dwelling Code,* this book has the latest technological advances in building design and construction. Among the changes are provisions for steel framing and energy savings. Also contains mechanical, fuel gas and plumbing provisions that coordinate with the *International Mechanical Code* and *International Plumbing Code.* **578 pages, 8¹/₂ x 11, $55.30. Also available on CD-ROM. $55.30**

Basic Plumbing with Illustrations, Revised

This completely-revised edition brings this comprehensive manual fully up-to-date with all the latest plumbing codes. It is the journeyman's and apprentice's guide to installing plumbing, piping, and fixtures in residential and light commercial buildings: how to select the right materials, lay out the job and do professional-quality plumbing work, use essential tools and materials, make repairs, maintain plumbing systems, install fixtures, and add to existing systems. Includes extensive study questions at the end of each chapter, and a section with all the correct answers. **384 pages, 8¹/₂ x 11, $33.00**

Residential Structure & Framing

With this easy-to-understand guide you'll learn how to calculate loads, size joists and beams, and tackle many common structural problems facing residential contractors. It covers cantilevered floors, complex roof structures, tall window walls, and seismic and wind bracing. Plus, you'll learn field-proven production techniques for advanced wall, floor, and roof framing with both dimensional and engineered lumber. You'll find information on sizing joists and beams, framing with wood I-joists, supporting oversized dormers, unequal-pitched roofs, coffered ceilings, and more. Fully illustrated with lots of photos. **272 pages, 8¹/₂ x 11, $34.95**

Moving to Commercial Construction

In commercial work, a single job can keep you and your crews busy for a year or more. The profit percentages are higher, but so is the risk involved. This book takes you step-by-step through the process of setting up a successful commercial business; finding work, estimating and bidding, value engineering, getting through the submittal and shop drawing process, keeping a stable work force, controlling costs, and promoting your business. Explains the design/build and partnering business concepts and their advantage over the competitive bid process. Includes sample letters, contracts, checklists and forms that you can use in your business, plus a CD-ROM with blank copies in several word-processing formats for both Mac and PC computers. **256 pages, 8¹/₂ x 11, $42.00**

Basic Engineering for Builders

If you've ever been stumped by an engineering problem on the job, yet wanted to avoid the expense of hiring a qualified engineer, you should have this book. Here you'll find engineering principles explained in non-technical language and practical methods for applying them on the job. With the help of this book you'll be able to understand engineering functions in the plans and how to meet the requirements, how to get permits issued without the help of an engineer, and  anticipate requirements for concrete, steel, wood and masonry. See why you sometimes have to hire an engineer and what you can undertake yourself: surveying, concrete, lumber loads and stresses, steel, masonry, plumbing, and HVAC systems. This book is designed to help the builder save money by understanding engineering principles that you can incorporate into the jobs you bid. **400 pages, 8¹/₂ x 11, $36.50**

Finish Carpentry: Efficient Techniques for Custom Interiors

Professional finish carpentry demands expert skills, precise tools, and a solid understanding of how to do the work. This new book explains how to install moldings, paneled walls and ceilings, and just about every aspect of interior trim — including doors and windows. Covers built-in bookshelves, coffered ceilings, and skylight wells and soffits, including paneled ceilings with decorative beams. **288 pages, 8¹/₂ x 11, $34.95**

How to Succeed With Your Own Construction Business

Everything you need to start your own construction business: setting up the paperwork, finding the work, advertising, using contracts, dealing with lenders, estimating, scheduling, finding and keeping good employees, keeping the books, and coping with success. If you're considering starting your own construction business, all the knowledge, tips, and blank forms you need are here. **336 pages, 8¹/₂ x 11, $28.50**

Construction Forms & Contracts

125 forms you can copy and use — or load into your computer (from the FREE disk enclosed). Then you can customize the forms to fit your company, fill them out, and print. Loads into Word for *Windows*™ , *Lotus 1-2-3, WordPerfect, Works,* or *Excel* programs. You'll find forms covering accounting, estimating, fieldwork, contracts, and general office. Each form comes with complete instructions on when to use it and how to fill it out. These forms were designed, tested and used by contractors, and will help keep your business organized, profitable and out of legal, accounting and collection troubles. Includes a CD-ROM for *Windows*™ and Mac. **400 pages, 8¹/₂ x 11, $41.75**

Basic Concrete Engineering for Builders

Basic concrete design principles in terms readily understood by anyone who has poured and finished site-cast structural concrete. Shows how structural engineers design concrete for buildings — foundations, slabs, columns, walls, girders, and more. Tells you what you need to know about admixtures, reinforcing, and methods of strengthening concrete, plus tips on field mixing, transit mix, pumping, and curing. Explains how to design forms for maximum strength and to prevent blow-outs, form and size slabs, beams, columns and girders, calculate the right size and reinforcing for foundations, figure loads and carrying capacities, design concrete walls, and more. Includes a CD-ROM with a limited version of an engineering software program to help you calculate beam, slab and column size and reinforcement. **256 pages, 8¹/₂ x 11, $39.50**

Contractor's Guide to QuickBooks Pro 2002

This user-friendly manual walks you through QuickBooks Pro's detailed setup procedure and explains step-by-step how to create a first-rate accounting system. You'll learn in days, rather than weeks, how to use QuickBooks Pro to get your contracting business organized, with simple, fast accounting procedures. On the CD included with the book you'll find a QuickBooks Pro file preconfigured for a construction company (you drag it over onto your computer and plug in your own  company's data). You'll also get a complete estimating program, including a database, and a job costing program that lets you export your estimates to QuickBooks Pro. It even includes many useful construction forms to use in your business: **328 pages, 8¹/₂ x 11, $46.50**

Also available: **Contractor's Guide to QuickBooks Pro 2001, $45.25**
Contractor's Guide to QuickBooks Pro 1999, $42.00

National Repair & Remodeling Estimator

The complete pricing guide for dwelling reconstruction costs. Reliable, specific data you can apply on every repair and remodeling job. Up-to-date material costs and labor figures based on thousands of jobs across the country. Provides recommended crew sizes; average production rates; exact material, equipment, and labor costs; a total unit cost and a total price including overhead and profit. Separate listings for high- and low-volume builders, so prices shown are specific for any size business. Estimating tips specific to repair and remodeling work to make your bids complete, realistic, and profitable. Includes a CD-ROM with an electronic version of the book with *National Estimator,* a stand-alone *Windows*™ estimating program, plus an interactive multimedia video that shows how to use the disk to compile construction cost estimates. **296 pages, 8¹/₂ x 11, $48.50. Revised annually**

Contractor's Plain-English Legal Guide

For today's contractors, legal problems are like snakes in the swamp - you might not see them, but you know they're there. This book tells you where the snakes are hiding and directs you to the safe path. With the directions in this easy-to-read handbook you're less likely to need a $200-an-hour lawyer. Includes simple directions for starting your business, writing contracts that cover just about any eventuality, collecting what's owed you, filing liens, protecting yourself from unethical subcontractors, and more. For about the price of 15 minutes in a lawyer's office, you'll have a guide that will make many of those visits unnecessary. **272 pages, 8¹/₂ x 11, $49.50**

Markup & Profit: A Contractor's Guide

In order to succeed in a construction business, you have to be able to price your jobs to cover all labor, material and overhead expenses, and make a decent profit. The problem is knowing what markup to use. You don't want to lose jobs because you charge too much, and you don't want to work for free because you've charged too little. If you know how to calculate markup, you can apply it to your job costs to find the right sales price for your work. This book gives you tried and tested formulas, with step-by-step instructions and easy-to-follow examples, so you can easily figure the markup that's right for your business. Includes a CD-ROM with forms and checklists for your use. **320 pages, 8¹/₂ x 11, $32.50**

Steel-Frame House Construction

Framing with steel has obvious advantages over wood, yet building with steel requires new skills that can present challenges to the wood builder. This new book explains the secrets of steel framing techniques for building homes, whether pre-engineered or built stick by stick. It shows you the techniques, the tools, the materials, and how you can make it happen. Includes hundreds of photos and illustrations, plus a CD-ROM with steel framing details, a database of steel materials and manhours, with an estimating program. **320 pages, 8¹/₂ x 11, $39.75**

Precision Framing

Here you'll find over 200 color photos and 40 detailed drawings along with the practical how-to information you need to frame a house. You'll find how to level a mudsill, plan and install center beams, planning and laying out exterior bearing walls, blocking, strapping, fire stops, radius openings and more. **218 pages, 8¹/₂ x 11, $29.95**

Roof Framing

Shows how to frame any type of roof in common use today, even if you've never framed a roof before. Includes using a pocket calculator to figure any common, hip, valley, or jack rafter length in seconds. Over 400 illustrations cover every measurement and every cut on each type of roof: gable, hip, Dutch, Tudor, gambrel, shed, gazebo, and more. **480 pages, 5¹/₂ x 8¹/₂, $22.00**

CD Estimator

If your computer has *Windows*™ and a CD-ROM drive, CD Estimator puts at your fingertips 85,000 construction costs for new construction, remodeling, renovation & insurance repair, electrical, plumbing, HVAC and painting. Quarterly cost updates are available at no charge on the Internet. You'll also have the *National Estimator* program — a stand-alone estimating program for *Windows*™ that Remodeling magazine called a "computer wiz," and Job Cost Wizard, a program that lets you export your estimates to QuickBooks Pro for actual job costing. A 40-minute interactive video teaches you how to use this CD-ROM to estimate construction costs. And to top it off, to help you create professional-looking estimates, the disk includes over 40 construction estimating and bidding forms in a format that's perfect for nearly any *Windows*™ word processing or spreadsheet program. **CD Estimator is $68.50**

National Construction Estimator

Current building costs for residential, commercial, and industrial construction. Estimated prices for every common building material. Provides manhours, recommended crew, and gives the labor cost for installation. Includes a CD-ROM with an electronic version of the book with *National Estimator*, a stand-alone *Windows*™ estimating program, plus an interactive multimedia video that shows how to use the disk to compile construction cost estimates. **616 pages, 8¹/₂ x 11, $47.50. Revised annually**

Rough Framing Carpentry

If you'd like to make good money working outdoors as a framer, this is the book for you. Here you'll find shortcuts to laying out studs; speed cutting blocks, trimmers and plates by eye; quickly building and blocking rake walls; installing ceiling backing, ceiling joists, and truss joists; cutting and assembling hip trusses and California fills; arches and drop ceilings — all with production line procedures that save you time and help you make more money. Over 100 on-the-job photos of how to do it right and what can go wrong. **304 pages, 8¹/₂ x 11, $26.50**

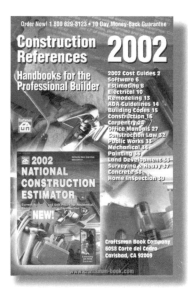

BUSINESS REPLY MAIL
FIRST CLASS MAIL PERMIT NO. 271 CARLSBAD, CA

POSTAGE WILL BE PAID BY ADDRESSEE

 Craftsman Book Company
6058 Corte del Cedro
P.O. Box 6500
Carlsbad, CA 92018-9974

BUSINESS REPLY MAIL
FIRST CLASS MAIL PERMIT NO. 271 CARLSBAD, CA

POSTAGE WILL BE PAID BY ADDRESSEE

 Craftsman Book Company
6058 Corte del Cedro
P.O. Box 6500
Carlsbad, CA 92018-9974

BUSINESS REPLY MAIL
FIRST CLASS MAIL PERMIT NO. 271 CARLSBAD, CA

POSTAGE WILL BE PAID BY ADDRESSEE

 Craftsman Book Company
6058 Corte del Cedro
P.O. Box 6500
Carlsbad, CA 92018-9974